W0112746

Chemical Process Dynamics and Controls

Chemical Process Dynamics and Controls

Contributors :
Yong Pan,
Liu Yang, *et al.*

AURIS REFERENCE LTD.
London, UK

Chemical Process Dynamics and Controls
Contributors : Yong Pan *and* Liu Yang, *et al.*

Auris Reference Ltd., UK

www.aurisreference.com

United Kingdom

Copyright 2016

Printed in 2017 for Sale in the Indian Subcontinent

The information in this book has been obtained from highly regarded resources. The copyrights for individual articles remain with the authors, as indicated. All chapters are distributed under the terms of the Creative Commons Attribution License, which permit unrestricted use, distribution, and reproduction in any medium, provided the original author and source are credited.

Notice

Contributors, whose names have been given on the book cover, are not associated with the Publisher. The editors and the Publisher have attempted to trace the copyright holders of all material reproduced in this publication and apologise to copyright holders if permission has not been obtained. If any copyright holder has not been acknowledged, please write to us so we may rectify.

Reasonable efforts have been made to publish reliable data. The views articulated in the chapters are those of the individual contributors, and not necessarily those of the editors or the Publisher. Editors and/or the Publisher are not responsible for the accuracy of the information in the published chapters or consequences from their use. The Publisher accepts no responsibility for any damage or grievance to individual(s) or property arising out of the use of any material(s), instruction(s), methods or thoughts in the book.

No part of this publication maybe reproduced, stored in a retrieval system or transmitted in any form or by any means, electronic, mechanical, photocopying, recording, scanning or otherwise without prior written permission of the publisher.

Chemical Process Dynamics and Controls

ISBN: 978-1-78154-510-2

British Library Cataloguing in Publication Data
A CIP record for this book is available from the British Library

Exclusively distributed by CBS Publishers & Distributors Pvt. Ltd.

Sales & Distribution Rights only for India, Pakistan, Bangladesh, Sri Lanka, Nepal and Bhutan.This book is not to be sold outside these territories.

PREFACE

Chemical Process dynamics is a field in which scientists study the rates and mechanisms of chemical reactions. It also involves the study of how energy is transferred among molecules as they undergo collisions in gas-phase or condensed-phase environments.

Therefore, the experimental and theoretical tools used to probe chemical dynamics must be capable of monitoring the chemical identity and energy content of the reacting species. Moreover, because the rates of chemical reactions and energy transfer are of utmost importance, these tools must be capable of doing so on time scales over which these processes, which are often very fast, take place.

It is a field within physical chemistry, studying why chemical reactions occur, how to predict their behavior, and how to control them. It is closely related to chemical kinetics, but is concerned with individual chemical events on atomic length scales and over very brief time periods. It considers state-to-state kinetics between reactant and product molecules in specific quantum states, and how energy is distributed between translational, vibrational, rotational, and electronic modes.

This page left intentionally blank.

CONTENTS

LIST OF CONTRIBUTORS

Yong Pan

State Key Laboratory of NBC Protection for Civilian, Research Institute of Chemical Defense, Yangfang, Changping District, Beijing 102205, China; E-Mails: Yangliujinjin@sina.com (L.Y.); Sdmuta@163.com (N.M.); Shaoshengyu@163.com (S.S.)

Liu Yang

State Key Laboratory of NBC Protection for Civilian, Research Institute of Chemical Defense, Yangfang, Changping District, Beijing 102205, China; E-Mails: Yangliujinjin@sina.com (L.Y.); Sdmuta@163.com (N.M.); Shaoshengyu@163.com (S.S.)

Ning Mu

State Key Laboratory of NBC Protection for Civilian, Research Institute of Chemical Defense, Yangfang, Changping District, Beijing 102205, China; E-Mails: Yangliujinjin@sina.com (L.Y.); Sdmuta@163.com (N.M.); Shaoshengyu@163.com (S.S.)

Shengyu Shao

State Key Laboratory of NBC Protection for Civilian, Research Institute of Chemical Defense, Yangfang, Changping District, Beijing 102205, China; E-Mails: Yangliujinjin@sina.com (L.Y.); Sdmuta@163.com (N.M.); Shaoshengyu@163.com (S.S.)

Wen Wang

Institute of Acoustic, Chinese Academy of Science, Zhongguancun Street, Haidian District, Beijing 100080, China; E-Mails: Wangwenwq@mail.ioa.ac.cn (W.W.); Xiexiao08@mails.ucas.ac.cn (X.X.); Heshitang@ mail.ioa.ac.cn (S.H.)

Xiao Xie

Institute of Acoustic, Chinese Academy of Science, Zhongguancun Street, Haidian District, Beijing 100080, China; E-Mails: Wangwenwq@mail.ioa.ac.cn (W.W.); Xiexiao08@mails.ucas.ac.cn (X.X.); Heshitang@ mail.ioa.ac.cn (S.H.)

Shitang He

Institute of Acoustic, Chinese Academy of Science, Zhongguancun Street, Haidian District, Beijing 100080, China; E-Mails: Wangwenwq@mail.ioa.ac.cn (W.W.); Xiexiao08@mails.ucas.ac.cn (X.X.); Heshitang@ mail.ioa.ac.cn (S.H.)

Chapter 1

INTRODUCTION TO PROCESS CONTROL

Process control is the study and application of automatic control in the field of chemical engineering. The primary objective of process control is to maintain a process at the desired operating conditions, safely and efficiently, while satisfying environmental and product quality requirements. Proper application of process control can actually improve the safety and profitability of a process. Even though rapidly decreasing costs of digital devices and increasing computers peed have enabled high-performance measurement and control systems, it is not an easy task to achieve this because modern plants tend to be difficult too per ate due to high complexity and highly integrated process units. The possibility of improving the performance and the profitability is illustrated by the figures below.

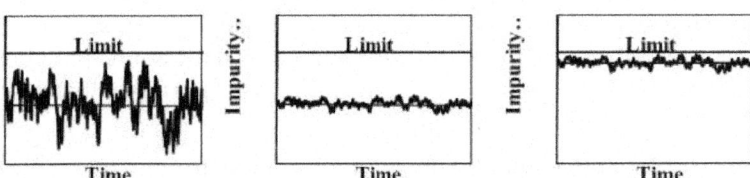

In this example, acceptable product quality requires that the impurity of the product is below the limit indicated in the figures. However, we do not want to make the product purer than necessary, because it would increase the production costs.

The figure to the left illustrates a situation where the impurity fluctuates a lot, but the quality requirements are fulfilled. The figure in the center illustrates that the fluctuations can be reduced by better control. Then it is possible to increase the average impurity in the product, as illustrated by the figure to the right, without violation of the quality requirements. Obviously, this also reduces the production costs.

As a consequence of global competition, rapidly changing economic conditions, and stringent environ-mental and safety regulations, process control has become increasingly important in the process industries. It is also clear that the scope and importance of process control technology will continue to expand. Consequently, *chemical engineers need to master this subject* in order to be able to design and operate modern plants.

PROCESS DYNAMICS

A process is a *dynamical system*, whose behaviour changes over time. Control systems are needed to handle such changes in the process. Thus, it is important to understand the *process dynamics* when a control system is designed. Mathematically, the process dynamics can be described by differential equations. Unsteady-state (or transient)process behaviour then corresponds to a situation, where (at least some) time derivatives of the differential equations are nonzero.

Transient operation occurs during important situations such as start-ups and shutdowns, unusual process disturbances, and planned transitions from one product grade to another. Even at normal operation, a process does not operate at a steady state (with all time derivatives of the differential equations exactly zero) because there are always variations in external variables, such as feed composition or cooling medium temperature. Thus, *knowledge of steady-state* (or static) *process properties*, taught in many engineering courses, *is not sufficient for control design.*

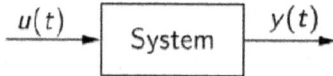

A dynamical system can be defined as a combination of components that act together to perform a certain objective. Conceptually, it is some isolated part of the universe that is of interest to us. For analysis and design purposes the full system of interest is usually decomposed into a number of subsystems that interact with each other. Such a subsystem (or even the full system) can be illustrated graphically by a block as shown in the figure.

Every (sub)system interacts with its environment through two groups of variables: input variables u(t) which affect the system behaviour in some way (*e.g.* a change in the feed composition), and output y(t)variables which give information about the system behaviour (*e.g.* a change in product quality). The argument "t" indicates that the values of the variables change over time.

Feedback Control

The study of process control introduces a major new concept: *feedback control.* This concept is central to most automation systems that monitor a process and adjusts some variables to maintain the system at (or near) desired conditions. Feedback is a topic studied in many engineering disciplines — for example, chemical, electrical, and mechanical engineering — and it is applied to a wide range of physical systems from electrical circuits to guided missiles and robots. Chemical engineers apply these principles to heat exchangers, mass transfer equipment, chemical reactors, and so forth.

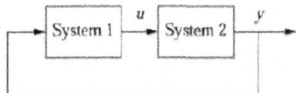

Feedback occurs when two (or more) dynamical systems are connected to-gether such that each system influences the other, as illustrated by the figure. Here, "System 1" could be a controller and "System 2" the process being controlled. Simple causal reasoning about a feedback system is difficult because the first system influences the second and the second system influences the first, leading to a circular argument. This makes reasoning based on cause and effect tricky, and it is necessary to analyze the system as a whole. A consequence of this is that the behaviour of feedback systems is often counter intuitive, and it is therefore necessary to use formal mathematical methods to understand them.

Feedback has many interesting and useful properties. It makes it possible to design systems that work in a desired way even though the subsystems are not exactly known. An unstable system can be stabilized using feedback, and the effects of external disturbances can be reduced. Feedback also offers process designers new degrees of freedom because it increases flexibility.

PROCESS CONTROL DEFINITIONS AND TERMINOLOGY

Process controls is a mixture between the statistics and engineering discipline that deals with the mechanism, architectures, and algorithms for controlling a process. Some examples of controlled processes are:

- Controlling the temperature of a water stream by controlling the amount of steam added to the shell of a heat exchanger.
- Operating a jacketed reactor isothermally by controlling the mixture of cold water and steam that flows through the jacket of a jacketed reactor.
- Maintaining a set ratio of reactants to be added to a reactor by controlling their flow rates.
- Controlling the height of fluid in a tank to ensure that it does not overflow.

To truly understand or solve a design problem it is necessary to understand the key concepts and general terminology. The paragraphs below provide a brief introduction to process controls as well as some terminology that will be useful in studying controls. As you begin to look at specific examples contained here, as well as elsewhere on the wiki, you will begin to gain a better grasp on how controls operate and function as well as their uses in industry.

Process Control Background

The role of process control has changed throughout the years and is continu-ously shaped by technology. The traditional role of process control in industrial operations was to contribute to safety, minimized environmental impact, and optimize processes by maintaining process variable near the desired values. Generally, anything that requires continous monitoring of an operation involve the role of a process engineer. In years past the monitoring of these processes was done at the unit and were maintained locally by operator and engineers. Today many chemical plant have gone to full automation which means that engineers and operators are helped by DCS that communicates with the instruments in the field.

Benefits of Process Control

The benefits of controlling or automating process are in a number of distinct area in the operation of a unit or chemical plant. Safety of workers and the community around a plant is probably concern number one or should be for most engineers as they begin to design their processes. Chemical plants have a great potential to do severe damage if something goes wrong and it is inherent the setup of process control to set boundaries on specific unit so that they don't injure or kill workers or individuals in the community.

The Objectives of Control

A control system is required to perform either one or both task:

1. **Maintain the process at the operational conditions and set points :** Many processes should work at steady state conditions or in a state in which it satisfies all the benefits for a company such as budget, yield, safety, and other quality objectives. In many real-life situations, a process may not always remain static under these conditions and therefore can cause substantial losses to the process. One of the ways a process can wander away from these conditions is by the system becoming unstable, meaning process variables oscillate from its physical boundaries over a limited time span. An example of this would be a water tank in a heating and cooling process without any drainage and is being constantly filled with water. The water level in the tank will continue to rise and eventually overflow. This uncontrolled system can be controlled simply by adding control valves and level sensors in the tank that can tell the engineer or technician the level of water in the tank. Another way a process can stray away from steady state conditions can be due to various changes in the environmental conditions, such as composition of a feed, temperature conditions, or flow rate.

2. **Transition the process from one operational condition to another:** In real-life situations, engineers may change the process operational conditions for a variety of different reasons, such as customer specifications or environment specifications. Although, transitioning a process from one operational condition to another can be detrimental to a process, it also can be beneficial depending on the company and consumer demands.

Examples of why a process may be moved from one operational set point to another:

1. Economics
2. Product specifications
3. Operational constraints
4. Environmental regulations
5. Consumer/Customer specifications
6. Environmental regulations
7. Safety precautions

Definitions and Terminology

In controlling a process there exist two type of classes of variables.

1. **Input Variable** – This variable shows the effect of the surroundings on the process. It normally refers to those factors that influence the process. An example of this would be the flow rate of the steam through a heat exchanger that would change the amount of energy put into the process. There are effects of the surrounding that are controllable and some that are not. These are broken down into two types of inputs.
   a. *Manipulated inputs:* variable in the surroundings can be control by an operator or the control system in place.
   b. *Disturbances:* inputs that can not be controlled by an operator or control system. There exist both measurable and immeasurable disturbances.
2. **Output variable-** Also known as the *control variable*These are the variables that are process outputs that effect the surroundings. An example of this would be the amount of CO_2 gas that comes out of a combustion reaction. These variables may or may not be measured.

   As we consider a controls problem. We are able to look at two major control structures.

1. Single input-Single Output (SISO)- for one control(output) varible there exist one manipulate (input) variable that is used to affect the process
2. Multiple input-multiple output(MIMO)- There are several control (output) variable that are affected by several manipulated (input) variables used in a given process.
   - **Cascade:** A control system with 2 or more controllers, a "Master" and "Slave" loop. The output of the "Master" controller is the setpoint for the "Slave" controller.
   - **Dead Time:** The amount of time it takes for a process to start changing after a disturbance in the system.
   - **Derivative Control:** The "D" part of a PID controller. With derivative action the controller output is proportional to the rate of change of the process variable or error.*

- **Error:** In process controls, error is defined as: Error = setpoint - process variable.
- **Integral Control:** The "I" part of a PID controller. With integral action the controller output is proportional to the amount and duration of the error signal.
- **PID Controller:** PID controllers are designed to eliminate the need for continuous operator attention. They are used to automatically adjust system variables to hold a process variable at a setpoint. Error is defined above as the difference between setpoint and process variable.
- **Proportional Control:** The "P" part of a PID controller. With proportional action the controller output is proportional to the amount of the error signal.
- **Setpoint:** The setpoint is where you would like a controlled process variable to be.

Design Methodology for Process Control

1. **Understand the process:** Before attempting to control a process it is necessary to understand how the process works and what it does.
2. **Identify the operating parameters:** Once the process is well understood, operating parameters such as temperatures, pressures, flow rates, and other variables specific to the process must be identified for its control.
3. **Identify the hazardous conditions:** In order to maintain a safe and hazard-free facility, variables that may cause safety concerns must be identified and may require additional control.
4. **Identify the measurables:** It is important to identify the measurables that correspond with the operating parameters in order to control the process.

   Measurables for process systems include:
   - Temperature
   - Pressure
   - Flow rate
   - pH
   - Humidity
   - Level
   - Concentration
   - Viscosity
   - Conductivity
   - Turbidity
   - Redox/potential
   - Electrical behaviour
   - Flammability

5. **Identify the points of measurement:** Once the measurables are identified, it is important locate where they will be measured so that the system can be accurately controlled.

6. **Select measurement methods:** Selecting the proper type of measurement device specific to the process will ensure that the most accurate, stable, and cost-effective method is chosen. There are several different signal types that can detect different things.

   These signal types include:

   • Electric
   • Pneumatic
   • Light
   • Radiowaves
   • Infrared (IR)
   • Nuclear

7. **Select control method:** In order to control the operating parameters, the proper control method is vital to control the process effectively. On/off is one control method and the other is continuous control. Continuous control involves Proportional (P), Integral (I), and Derivative (D) methods or some combination of those three.

8. **Select control system:** Choosing between a local or distributed control system that fits well with the process effects both the cost and efficacy of the overall control.

9. **Set control limits:** Understanding the operating parameters allows the ability to define the limits of the measurable parameters in the control system.

10. **Define control logic:** Choosing between feed-forward, feed-backward, cascade, ratio, or other control logic is a necessary decision based on the specific design and safety parameters of the system.

11. **Create a redundancy system:** Even the best control system will have failure points; therefore it is important to design a redundancy system to avoid catastrophic failures by having back-up controls in place.

12. **Define a fail-safe:** Fail-safes allow a system to return to a safe state after a breakdown of the control. This fail-safe allows the process to avoid hazardous conditions that may otherwise occur.

13. **Set lead/lag criteria:** Depending on the control logic used in the process, there may be lag times associated with the measurement of the operating parameters. Setting lead/lag times compensates for this effect and allow for accurate control.

14. **Investigate effects of changes before/after:** By investigating changes made by implementing the control system, unforeseen problems can be identified and corrected before they create hazardous conditions in the facility.

15. **Integrate and test with other systems:** The proper integration of a new control system with existing process systems avoids conflicts between multiple systems.

DIGITAL CONTROL SYSTEMS

Digital Control Systems(DCS) also known as Distributed Control System is the brain of the control system. It is used mainly for the automation of a manufacturing process and manages the logic that exist for major unit operations. A DCS in the past was tailor made for the process, plant or company that intended to use the structure to control and model it's process. Before the beginning of the DCS era there were pneumatic devices that controlled process and engineers manually turned valves on the site. Modeling of the systems was made possible by DCS as it allowed the ability to record and manage process from comfort of a computer screen. Because of DCS we are able to control processes remotely and gain a better understanding of how the process operate and how they can be improved to both increase safety and increase profit possibilities.

Control Systems are collectively named as "ICSS" Integrated Control and Safety System. Distinctly identified as "BPCS" Basic Process Control System."SIS" Safety Instrumentation System."F&G" Fire and Gas System.

How Does a DCS Work?

In the field you have sensors and gauges that give and recieveinformation. They convert this information into a electric signal that is sent to a control room somewhere in the field. This control room has programmed logic that is able to converts the signal into a pressure, flow rate, concentration, temperature, or level. This logic also contains the information that controls the process and takes the signal compares it with the set point sent from the operator may or may not be in the field and sends a signal to the manipulated variables in the field. The DCS covers all of the computer logic from the operator screen to the field box that contain the logic.

Shutdown Systems

Shutdown system are the emergency setting of the logic to make sure the process can be contained and is environmentally safe. These setting are important for emergency response of the system. It is the job of the DCS to contain the logic for the shutdown system and be able to operate when a process exceed a certain limit.

CURRENT SIGNIFICANCE

Industrial processes are central to the chemical engineering discipline. Generally, processes are controlled in order to do things such as maximize safety, minimize cost, or limit effects on the environment. This course aims to help undergraduate engineering students understand the mechanisms used to moderate these processes, such as to control their output.

Automation

Generally, process controls are designed to be automated. This means that given a change in system response, the control system can act on its own to account for it. In order to minimize cost, automated systems have become wide-spread throughout industry. Before automation, a huge amount of labor would be required to run even the simplest processes. For example, a technician might be hired to monitor the temperature in a reaction vessel, and operate a valve to manipulate the cooling water flow rate in the jacket. In a sense, this technician operated as a control system. If the temperature reading is too high, the technician will manipulate the system in order to bring the temperature down. Via automation, this simple, arduous labor can be done by an algorithm.

By designing an effective control system, even the most complicated of processes can be run with minimal worker supervision. Telephone operators, for example, have largely been replaced by automated telephone switch boards. Removing the need for telephone operators decreases operating cost for phone companies, thereby allowing the general consumer to pay less for phone service. Automated process controls, therefore, are enormously important in the modern world.

Failures in Process Control

Process controls can have a huge impact on surrounding communities, as well as the environment. An engineer of a large-scale process, therefore, has an important ethical responsibility to operate a process safely and properly. These responsibilities extend well beyond the scope of merely the company for which they work. Catastrophic failures in process control remind us of the importance of control systems and engineering in today's world.

Bhopal, India Disaster

The Bhopal Gas Tragedy in Bhopal, India on December 3, 1984 was a large toxic gas leak that killed thousands of people in the surrounding area. A tank with 42 tons of methyl isocyanate(MIC) was contaminated with water. This in turn caused a run away reaction that greatly increased the pressure and temperatures in the tank, which forced the emergency venting of the toxic gases to the atmosphere.

This tragedy was largely due to the failure or lack of safety controls:

1. Runaway reaction as temperature and pressure increased without regulation
2. MIC was suppose to be cooled, however in the Bhopal plant the refrigeration system was not turned on. Temperature control on the tank could have greatly hindered the runaway reaction that ensued with the addition of water.
3. Flare tower to handle the leakage of toxic gases was not functional
4. The plant also had vent scrubbers, which were also not functional
5. Water curtain, which would neutralize some escaping gas, not designed properly. It was not tall enough to reach the top of the flare tower, making it essentially worthless.

6. Alarms that would have alerted to a malfunction in the tank had not been operational for 4 years

The figure below illustrates some of these failures:

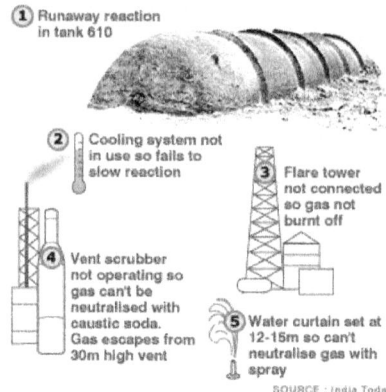

SOURCE : India Today

Had at least some of these been functioning the amount of toxic gas released would have been substantially reduced.

Results

From this tragedy we can see that if the plant had proper safety controls the effects of the disaster would have been greatly reduced. Therefore as a chemical engineer it is our responsibility to society to provide sufficient safety controls to chemical processes in order to prevent disasters such as the Bhopal Gas Tragedy from happening. Unfortunately, industrial negligence is still a problem in many third-world countries.

Three Mile Island Disaster

Overview

One of the largest and most far reaching plant failures in United States history took place at a nuclear power plant on Three Mile Island in March 1979. The event was caused by either a mechanical or electrical failure of the main feed water pumps causing the power plant to begin to overheat. As the heat increased, the control scheme caused the turbine and reactor to shut down. This caused a pressure increase in the primary system (nuclear portion of the plant) and a relief valve automatically opened to release some of the pressure to prevent the reactor from blowing.

All of these actions were well designed to prevent a significant event from happening. The problem was that the release valve did not close properly when the pressure in the reactor was relieved. As a result, when the reactor started back up, coolant in the core of the reactor was lost through the pressure relief valve. Because there was no control mechanism that measured the level of the coolant in the reactor, the operators, who only judged the water level by the pressure in the reactor, actually decreased coolant flow to the reactor.

The figure below is a simplified diagram of the TMI-2-plant:

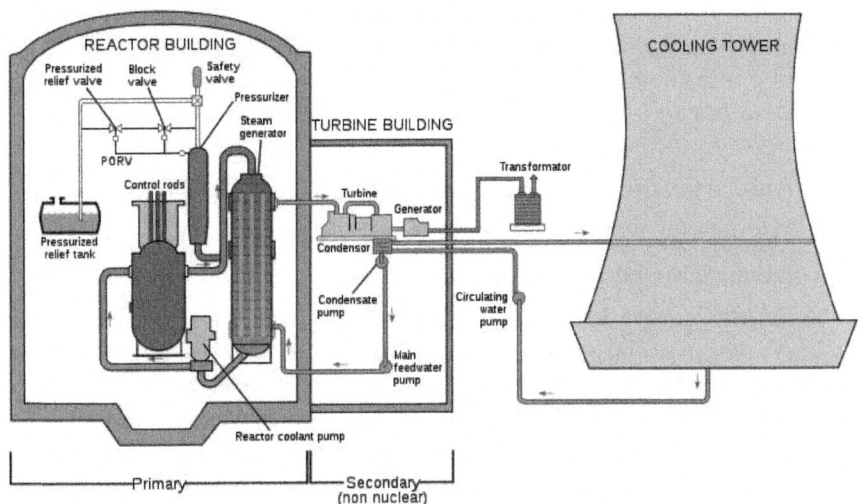

The result of the control design failure that prevented the operators from cooling the reactor was that the rods that held the nuclear fuel melted causing the fuel to also melt. This is the worst thing to have happen in a nuclear power plant and is what happened to cause the disaster at Chernobyl. Thankfully, the accident was largely contained and although the entire nation watched for 3 days as the threat of an explosion or breach of containment loomed, 0 deaths or injuries resulted. In fact, corrective steps were so successful that the average increase in radiation to the surrounding population was around 1% and the maximum increase at the boundary to the site is estimated to be less than 100% of the natural background radiation present in the region.

Results

The accident at Three Mile Island showed the importance of proper design of control systems. As a result the US Nuclear Regulatory Commission took steps to tighten their regulation and increase the safety requirements on Nuclear Power Plants. These included revamping operator training as well as increasing the design and equipment requirements. This also brought the dangers of all industrial processes to the forefront and reminded people of the importance of the safety of the communities surrounding chemical and power plants.

Unfortunately, the incident also inspired intense fear of nuclear power in the general population and is partially responsible for the reduced build rate for new nuclear power plants since that time. Although control failures can be corrected fairly quickly, after one safety issue it is difficult to convince the general public that engineers have fixed the problem and that it will not happen again.

VERBAL MODELING

Every process requires a great deal of planning in order to successfully accomplish the goals laid out by its designers and operators. In order to accomplish these goals, however, personnel that are not familiar with the design must fully understand the process and the functions of the control systems. Control systems consist of equipment (measuring devices, valves, *etc.* and human intervention (plant operators and designers). Control systems are used to satisfy three basic needs of every process:

1. Reduce the influence of external disturbances
2. Promote the stability of the process
3. Enhance the performance of the process

Verbal modeling is used for creating and also understanding a process control system. Verbal modeling consists of first receiving and then gathering information about the process. A step-by-step process is then used to describe the control systems used to satisfy constraints and objectives that have been outlined. In the following sections you will read what requirements are generally outlined for process control and the step-by-step method used to meet these requirements.

Prerequisite Information Regarding a Process

For the sake of this article it is assumed that a process has already been designed and that certain restraints and criteria are provided by either a customer, management, or the government. The goal of this chapter is to classify the types of criteria that are usually given. These criteria then become the conditions that the control systems employed must satisfy. In general, there will be five sets of criteria, often coming from different people and institutions. By gathering all of these criteria you will be able to describe the control system. If you do not have a complete list of these criteria you must research the process to determine these constraints before beginning the step-by-step process below.

Safety

The safe operation of a process is the biggest concern of those working in the plant and those that live in the surrounding community. The temperatures, pressures, and concentrations within the system should all fall within acceptable limits, and these limits can be dictated by either government agencies or company policy.

Production Objectives

The production objectives usually include both the amount and purity of the desired product. This criterion is generally set by the company or customer.

Environmental Regulations

These come in the form of restrictions on the temperature, concentration of chemicals, and flow rate of streams exiting a plant. State and federal laws, for instance, may dictate the exit temperature of a cooling water stream into a lake in order to prevent harm to aquatic wildlife.

Operational Constraints

Equipment found in the plant may have their own unique limitations, such as temperature or pressure that require proper control and monitoring. For instance, a thermocouple may be damaged at extremely high temperatures, thus the location of the thermocouple must be accounted for.

Economics

In general, a company will operate so that its profits are maximized. The process conditions that maximize these profits are determined by way of optimization. Many costs must be considered when optimizing process conditions. Some of these costs are fixed, or will not change with process variables (*i.e.* equipment costs) and others are variable, or do depend on process variables (*i.e.* energy costs). The overall process is usually limited by certain factors including availability of raw materials and market demand for the final product. Therefore, the economics of a process must be well understood before process changes are enforced.

Step-by-Step Method For Describing Controls and Their Purpose

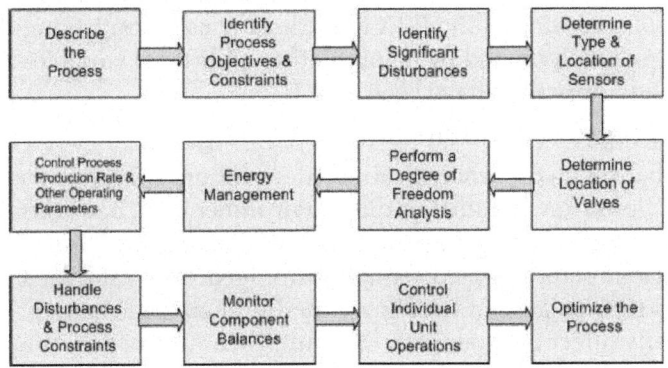

1. Describe the Process

A brief description of the general process is needed while not dwelling on the details and calculations involved. The major steps of the process, as well as inputs and outputs of the process, should be stated. A simple diagram should be provided detailing the chemical process to help visualize the process.

2. Identify Process Objectives and Constraints

The objectives and constraints of the process must be identified before process control actions can be performed.

The process objectives include the type, quantity, and quality of the product that is to be produced from the process. The economic objectives, such as the desired levels of raw material usage, costs of energy, costs of reactants, and price of products, should also be identified.

The process constraints include three different categories: operational, safety, and environmental limitations. *Operational constraints* refer to the limits of the equipment used in the process. For instance, a liquid storage tank can only hold a certain volume. *Safety constraints* describe the limits when the people or the equipment may be in danger. An example would be a pressure limitation on a reactor, which if exceeded, could result in an explosion. *Environmental constraints* limit how the process can affect the immediate surroundings. For example the amount of harmful chemicals that can be released before damage is done to nearby water supplies. All of these constraints should be mentioned to build a robust control system.

Careful reading of the information provided to you by the customer, management, and government is required in order to properly identify each constraint and objective. Often times, the process objectives will be very clearly laid out by the needs of the customer or management. Operational constraints, or the limitations of the equipment being used, must be researched for each piece of equipment used in the process. Generally, by satisfying the operational constraints a good portion of safety constraints are satisfied as well, but additional safety constraints may exist and must be investigated by researching company policy and governmental regulations. Environmental regulations also have to be researched through resources such as the EPA and Clean Air Act. Satisfying the economic aspect is largely determined by manipulating additional variables after all other constraints and objectives have been met.

3. Identify Significant Disturbances

Disturbances, in the sense of process description, are defined as inputs or external conditions from the surrounding environment that have certain properties that cannot be controlled by the plant personnel. Examples of disturbances include ambient air temperature, feed temperature, feed flow rate, feed composition, steam pressure changes, and cooling water temperature changes. Disturbances can drastically affect the operation of a unit. A control system should be able to effectively handle all process disturbances. As such, all possible disturbances must be identified and these disturbances need to be accounted for by the development of contingency plans within the process.

4. Determine Type and Location of Sensors

A proper design must ensure that adequate measurements of the system are obtained to monitor the process. To meet this goal, sensors must be chosen to accurately, reliably, and promptly measure system parameters. Such parameters include temperature, flow rate, composition, and pressure. Placement of sensors is important both in the usefulness of measurements as well as the cost of the system. Sensors should be placed such that the measured quantities are appropriate in addressing control objectives.

5. Determine the Location of Control Valves

Valves must be placed in a location to control variables that impact the control objectives. For example, control of the temperature of a reactor could be obtained by placing a valve on either the stream of heating/ cooling fluids or by placing a valve on the feed stream to the reactor. One must determine which streams should be manipulated to meet process objectives.

6. Perform a Degree of Freedom Analysis

The degrees of freedom in a system are equal to the number of manipulated streams (determined in step 5) minus the number of control objectives and control restraints (determined in step 2). A degree of freedom analysis is used to determine if a system is being under- or over-specified by the process objectives. The degrees of freedom come from the number of knowns and unknowns that are specified within the system. If there are extra degrees of freedom present in a system, unused manipulated variables can be used to optimize the process. If there are negative degrees of freedom, a system is over-specified because more objectives and restraints exist than manipulated streams. In this case, all objectives cannot necessarily be met simultaneously and the least important objectives must be neglected. A system with zero degrees of freedom is fully specified. All objectives can be met, but there is no room for optimization.

7. Energy Management

In any system with exothermic or endothermic reactions, distillation columns, or heat exchangers, energy management becomes a factor that must be accounted for. Heat must be removed from exothermic reactions in order to prevent reactor runaway, and heat must be supplied to endothermic reactions to ensure desired production rates. Strategies such as pre-heating feed streams with the excess heat from a product stream are helpful in maintaining efficient usage of energy, however, they also result in more complex processes that may require more intricate control systems.

8. Control Process Production Rate and Other Operating Parameters

The production rate can be controlled by a variety of manipulated variables. One manipulated variable may be the feed rate. The plant feed rate can be changed and each subsequent unit can use its controls to accommodate this change, ultimately resulting in a change in the final production rate. Other manipulated variables may also include reactor conditions, such as temperature and pressure. Temperature and pressure affect reaction rates and can be used to alter the final production rate. It is important to choose the most suitable manipulated variable to control production rate.

In addition to the production rate, other control objectives must be effectively managed by manipulated variables. For example, temperature of an exothermic reactor may be controlled by the flow of a coolant stream passing over it in order to avoid dangerous high temperatures. The pressure of a reactor may be controlled by the flow of feed gas in order to comply with the pressure limitations of the vessel.

9. **Handle Disturbances and Process Constraints**

The effects of disturbances should be minimized as much as possible, in order to maintain the system at desired conditions and meet all process objectives and constraints. Feedback or feedforward are specific control techniques and are common ways to overcome disturbances. A feedback control works by studying the downstream data and then altering the upstream process. The actions executed are reactive. Feedback can be viewed as an if-then statement: if a feed's temperature is detected to be lower than desired, then steam can be used to preheat the feed. Feedforward is a more proactive approach in that it adjusts a manipulated variable before the disturbance is felt in the process. Hence, if a sensor indicates low temperatures upstream of the feed, the feedforward control will counteract the effect of the cooler upstream temperatures by preheating the feed before the feed temperature is effected. Note that a disturbance must be detectable and measurable in order for the feedforward control to fix the anticipated disturbance before the system is effected.

Additionally, if constraints are reached during the process, controls should be implemented to avoid safety, operational, or environmental hazards. This can also be done with feedback and feedforward controls on manipulated variables.

10. **Monitor Component Balances**

Every component within a process, whether it is inert or not, should be accounted for at every step of the system in order to prevent accumulation. This step is more crucial in processes that involve recycle streams. If such a stream is present, a purge stream is often necessary to remove unwanted components. In addition, component balances are used to monitor yield and conversion or reveal locations in the process where loss may be occurring. In order to monitor component balances, composition sensors are used.

11. **Control Individual Unit Operations**

Most systems used today in industry employ the use of multiple unit operations. Each of these unit operations, however, needs to be fully controllable in the sense that it has a control system that can adjust manipulated variables in order to maintain other parameters. For instance, if an absorber is present, the system must be able to control the liquid solvent feed as some ratio to the gas feed. Another example is a crystallizer. The refrigeration load of the crystallizer must be controllable in order to control the temperature.

12. **Optimize the Process**

In most cases, there will be certain aspects of a process that will not be dictated to a designer and can be changed to make the overall process more economical for the company. These are referred to as "unaccounted for" degrees of freedom and can be implemented as new control valves or adjustable controller setpoints.

Alternative Method of Verbal Modeling

This method can easily be applied to describe an entire system of unit processes.

1. Describe the process in words

Some of the important questions to answer before delving deeper into a model are:

- What are the components entering the system?
- How do they enter? Separately? Combined stream? What physical states are they in?
- What happens inside the unit process and what comes out at each exit point?

Remember to keep this part simple. There is no need to include chemical formulations or equations of any sort. Just lay out the basic flow of material.

2. Define the primary goal of the process

The primary goal should be simple. Often, it is to maintain a specific measured variable above a minimum or below a maximum. In this step, the only thing that needs to be determined is what the main goal is, and a few supporting details about why this is an important goal to achieve.

For example, a primary goal could be to minimize the concentration of Compound Y in orange juice because studies show Compound Y gives the juice a bad aftertaste.

3. Identify secondary processes that influence the primary goal

In a typical unit process, the primary goal will be directly influenced by one or two other aspects of the system. These can include temperature, pressure, inlet conditions, and more and can occur at various points in the process.

The goal of this step is to determine which of these other process variables will be most likely to influence the primary goal and to step down from there.

For example, the temperature of the orange juice mixer could have the greatest influence on production of Compound Y.

4. Identify safety and environmental risks

Next, you need to identify all of the points in the process that represent any type of risk. This will be important later in determining which system variables need to be monitored.

Step through your process and identify any points that pose a significant risk of the hazards shown in the following figure.

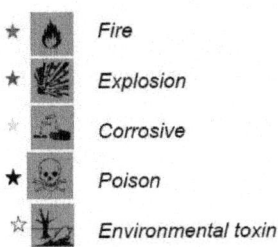

Examples include: Boilers represent fire and explosion risks. Any stream with a dangerous chemical can represent corrosive, poison, environmental, or all three risks.

5. Identify major costs associated with the process

How much something costs to produce is obviously a big deal in manufacturing. Identifying the largest sources of cost is critical in finding ways to reduce cost overall. Typical places to start identifying costs are at inlet streams (what is the cost of raw materials) and at any portion of the process where heat is added or removed.

It is important to include the high costs that can be associated with the risks identified in Step 4. Often the high cost of failure and risk exposure will determine what other seemingly costly steps must be taken to ensure the safety of the process.

6. Identify variables you can directly manipulate

The basics of the process have been laid out, and now it's important to determine what variables you can actually control. Typically, you only have direct control over the simplest of variables: switches and valves. Essentially, this just means that you cannot, in fact, choose a temperature for your system and implement it. What you can do, is control a valve or switch that activates heating or cooling to control the temperature.

During this step, you must decide where it is important to place these valves and/or switches. Use the information acquired previously about the primary goal and secondary effects to determine what variables are worth controlling. Remember that you don't need to put valves EVERYWHERE! Control valves are not costless and can also add unwanted complexity to your system. If you need to isolate equipment, you can install manual valves. Keep the control valves to the needed level.

7. Identify sources of variation

In order to write a control scheme, you need to know what values in your system will change and why. Some common causes of variation include:

- Environment: ambient temperature
- Other processes upstream or downstream: variable inlet conditions or outlet demand
- Economic forces: product worth, material costs
- Operators

Identifying what aspects of your process can be affected by these forces will allow you to assemble a more complete control scheme.

8. Describe your control system in words

Before you start trying to write everything out in computer code and mathematical equations, take the time to lay out your controls in words. This is much like preparing an outline before writing a paper. It can save you from many headaches later on.

One example of generic, simple syntax for verbal modeling is: Maintain [system variable] at specified level by adjusting [variable I can control].

The Barkel Method of Verbal Modeling

1. Understand the Process

Before you can control anything, you have to understand the process and how different parts of the system interact with each other. Make sure that the overall process is understood. This includes inputs and outputs as well as major steps, however specifics are not necessary. You should also be able to construct a diagram to help explain the process.

2. Identify Operating Parameters

Operating parameters can include temperature, pressure, flow, level, *etc.* Choose the parameters to manipulate in your system that will safely result in the desired output.

3. Identify Hazardous Conditions

Consider all possible dangerous aspects of your process when designing your system. This could include a chemical overheating or a vessel overflowing. It is imperative that you ensure the safety of your operators and being aware of all hazardous conditions can aid in this.

4. Identify Measurables

The main three measureables addressed in this class are: temperature, pressure, and flow. However, there are many more measureables, some more common than others. Here are some more: pH, humidity, level, concentration, viscosity, conductivity, turbidity, redox/potential, electrical behaviour, and flammability.

5. Identify Points of Measurement

It is important to place sensors in locations so that efficient and reliable measurements are taken of the system to monitor the process. For example, in a distillation column, temperature sensors will display different temperatures at different locations down the tower. The sensors must be positioned so that accurate readings of the system are given. Also, it is necessary to place sensors in an area of constant phase.

6. Measurement Methods (Thermo Couple? Choose for Range)

After identifying what is to be measured and where it will be measured, you have to decide how it will be measured. For example, a thermocouple can be used to measure the temperature. When choosing the equipment, be sure to check that the conditions of use fall within the recommended range of operation for the equipment.

7. Select Control Methods

Decide whether to use feedback, feed-forward, cascade, or other types of control methods.

8. Select Control System

A control system is a set of devices that will manipulate the actions of other devices in the system. A control system ranges from having an operator manually open and close a valve to running a system with feedback such as with PID controllers. Control systems can vary from relatively cheap to expensive. When picking a system, it would be most economic to choose the cheapest one that gets the job done. An example of a control system is a PIC, or programmable interface controller.

9. Select Control Limits

When choosing setpoints for controllers, you will also have to decide on a range that the values are allowed to fluctuate between before a corrective change is made. When selecting these limits, keep in mind that "equal" and "zero" do not exist due to the infinite number of decimals that electronics are now able to handle. So you must further define what "equal" means, when is it "close enough" to count as "equal" And when is the number small enough to count as "zero", is 0.1 or 0.01 or 0.0000000001 count as "zero"?

10. Define Control Logic

As every process is different, a customized code for each process must be written to tell the system what to do. For example, when a level control is a tank has reached a critically high point, the logic should spell out the necessary changes needed to bring the tank level back down. For example, this could be partially closing a valve upstream of the tank or partially opening a valve downstream of the tank.

11. Create Redundancy System

In the real world, you must balance cost and efficiency/safety. On one hand, you don't want an out-of-control system if one control fails. But on the other hand, you can't afford to order two of everything. The critical point to keep in mind is to optimize the safety while minimizing the cost.

12. Define "Fail-Safe"

A fail safe is a set up in the control logic to ensure that in the event of a failure of a control method, the system will automatically reach a safe condition so that there is little to no harm done to other equipment or personnel.

13. Set Lead/Lag Criteria

Valves and other equipment do not necessarily open/close or turn on/off at the exact instant a button is pressed or the control logic kicks in. There is often lag time associated with each controller. You must determine how long this lag time is so that you can take it into account.

14. Investigate Effects of Change Before/After

Be sure to investigate effects of changing each controller. For example, what are the effects of closing/opening this valve?

15. Integrate All Systems

Ensure that all systems are working together and that there are no holes in the system. Make sure that information does not fall through any cracks in the system.

[Note - we can use an example of the Barkel method - RZ]

COMMON ERRORS

1. **Impossible direct manipulations** *e.g.,* Change the concentration of salt in a tank

2. **Missing the forest for the trees** *e.g.,* Sacrificing product quality for tight level control on a tank

3. **Excessive or insufficient control** *e.g.,* Control every variable because you can or ignore the possibility of significant disturbances

Worked out Example 1

All Values in the Following Problems are Ficticious but Meant to be Logical

A heat exchanger uses steam to heat a stream of water from 50°F to 80°F. The water enters at a flow rate of 20 gallons per minute from a nearby lake. The process costs $65 per hour and yields a profit of $2 per gallon of product. The steam is provided by the plant and is 1000°F, which is a temperature that the pipes can sustain. For safety reasons, the exchanger may only run for 12 consecutive hours and requires 4 hours to cool down. Using more than 10,000 gallons of water per hour would cause an environmental disturbance to the water source. The diagram is shown below. Verbally model this system.

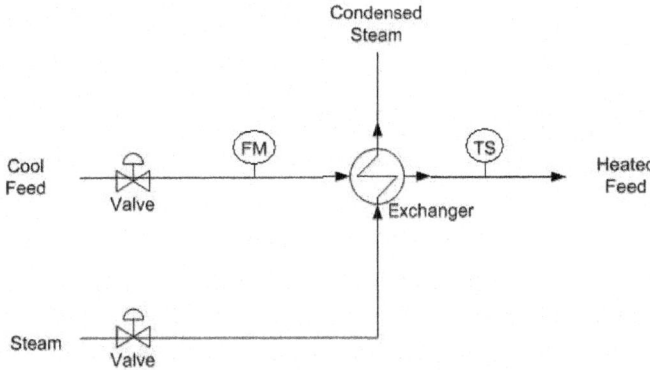

Solution

1. Describe the Process

The purpose of the process is to heat an incoming stream of water from a temperature of 50°F to a temperature of 80°F. The main equipment involved is a shell-and-tube heat exchanger.

2. Identify Process Objectives and Constraints

The product specification of the process is water at a flow of 20 gallons per minute and a temperature of 80°F.

Economically, the process costs $65 per hour to operate. There are no costs for the raw materials, as the only inputs to the system are water and steam. The finished product produces a profit of $2 per gallon. The economic objective is to reduce process costs while producing sufficient product.

The operational constraints and safety concerns are due to the pipes. The pipes can only sustain a temperature of 1000°F. Safety is a concern because attempting to heat the incoming water to a certain temperature may cause the heat exchanger to malfunction, leading to equipment damage and possible burn injuries to nearby personnel. The system may only operate for 12 consecutive hours, after which the system will need to be cooled down for 4 hours to avoid the aforementioned hazards. A simplified assumption is that there are no constraints on steam because it is provided by the plant and causes no safety issues. The only environmental constraints involve the incoming water stream. The incoming water is gathered from the nearby lake, and a stream of greater than 10000 gallons per hour would cause a disturbance in the equilibrium of the lake.

3. Identify Significant Disturbances

Significant disturbances can be found in the ambient air temperature, variable flow rates of the feed, and the temperature of the steam.

4. Determine the Type and Location of Sensors

A flow sensor (FM) is placed at the incoming water stream. A temperature sensor (TS) is located on the product water stream. A flow sensor is not needed for the steam stream for this problem because this value is not needed for control. A sensor could be placed here but the information is not needed for this problem.

5. Determine the Location of Control Valves

A flow valve is placed at the entrance of the incoming water stream. A flow valve is placed at the entrance of the steam.

6. Perform a Degree-of-Freedom Analysis

There are two manipulated variables: the flow of the water feed stream and the flow of the incoming steam. There are two control objectives: the flow of the feed stream, monitored by the flow sensor, and the temperature of the product, monitored by the temperature sensor. Therefore the system has zero degrees of freedom.

7. Energy Management

The incoming steam is used to transfer heat to the cool water feed. The temperature sensor on the product stream determines the applicable setting on the steam flow valve.

8. Control Process Production Rate and Other Operating Parameters

The process production rate is controlled by the flow valve on the entering water stream. The water temperature is controlled by the flow valve on the incoming steam.

9. Handle Disturbances and Process Constraints

Changes in the ambient air temperature can be detected by the temperature sensor, and can be corrected by the flow valve on the incoming steam stream. Variable flow rates of the water feed stream can be detected by the flow sensor and compensated by adjustments on the flow valve on the water feed stream. Changes in the temperature of the steam can be detected by the temperature sensor. The flow valve on the steam stream can be adjusted to increase or decrease the flow of steam and subsequently the amount of heat exchanged.

10. Monitor Component Balances

A vent is located on the heat exchanger to release excess steam from the system. Aside from that, any accumulation is unlikely and can be neglected.

11. Control Individual Unit Operations

The outlet temperature of the product stream is controlled by the flow valve on the steam feed stream. The flow of the incoming water stream is controlled by the flow valve on the water feed stream.

12. Optimize the Process

One might notice that the process is only using 1,200 gal/hr of water, well below the 10,000 gal/hr environmental constraint. If the profit of the process is linear with the flow-rate of water, then increasing the flow-rate of water will increase the profits for the company. (With the constraints specified, this is a Linear Programming optimization problem. The optimal setpoint falls on a boundary condition. However, the flow-rate of water entering the system is already specified, which results in zero degrees of freedom. (Zero degrees of freedom implies there are no further control valves or setpoints. Further investigation should be conducted to determine the reason for the flow-rate specification. When considering increasing the flow-rate of water into the system, one should also check that the other constraints are not violated.

Worked out Example 2

A process converts phenol into salicylic acid through a series of two reactors. Phenol and NaOH are fed in the liquid phase into the first reactor where it reacts with gaseous carbon dioxide that is pumped in. Assume constant fresh feed temperature and that the feed flow rate is within operational constraints. Management has dictated that salicylic acid production must be 200 moles per hour. Also, management would like the product stream to have a molar composition of 80% salicylic acid. Due to environmental concerns, a maximum flow rate of 10000 gallons per hour of cold water can be used to cool the first reaction chamber. The valve controlling the flow of cold water does not allow a flow rate in excess of 7500 gallons of water per hour. The salicylic acid product is used in another process to produce aspirin, which has a market value of $10 per mole. The first reactor can be operated at pressures up to 200atm, while the second can be operated at pressures up to 10 atm. The first reaction is exothermic, while the second reaction is assumed to generate negligible heat. A diagram of this process is shown below, as well as the reaction scheme. Verbally model this system.

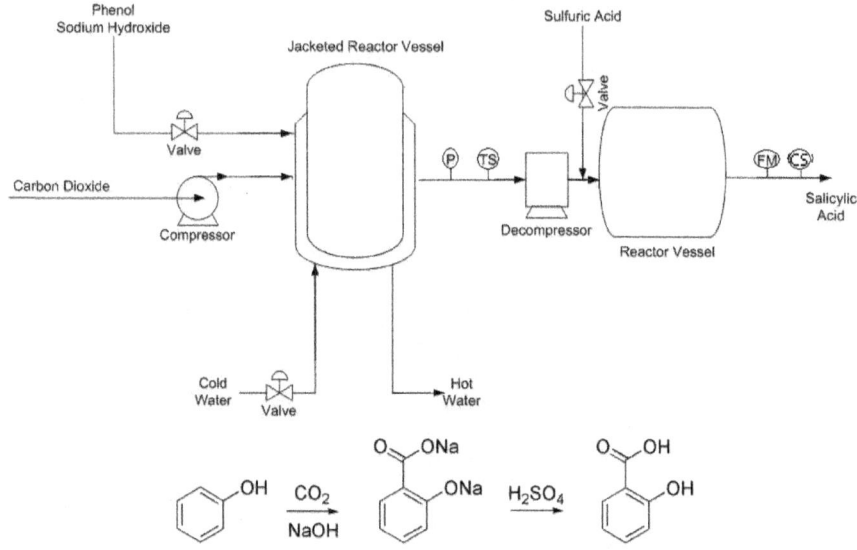

Solution

1. Describe the Process

The purpose of the process is to convert phenol into salicylic acid, the precursor for aspirin. First, the phenol is reacted with gaseous carbon dioxide (CO_2) and sodium hydroxide (NaOH) under high pressure and temperature in the first reactor. The product of this reaction is then combined with sulfuric acid to form salicylic acid in the second reactor. The reaction scheme is shown above.

2. Identify Process Objectives and Constraints

The process is expected to produce 200 moles per hour of salicylic acid. The product stream must contain at least 80% by moles of salicylic acid. The equipment used in the process dictates the operational limitations. The first reactor vessel can be operated up to a pressure of 200atm, while the second reactor vessel has a 10atm upward pressure limit. As such, pressures in excess of these limits must be avoided. Since the first reactor will generate a significant amount of heat, the heat must be removed to avoid damage to equipment and possible runaway reactions. Therefore, a heat exchanger (in the form of a reactor jacket in this case) with cool water should be included to decrease the temperature of the reactor. Economic concerns demand that phenol, sodium hydroxide, and sulfuric acid should not be used in extreme excess. The costs of these materials and the energy costs required to process them affect the overall profitability, so these compounds should not be wasted. Environmental regulations limit the use of water to cool the reactor at 10000 gallons per hour, however the valve constraints limits the amount of water to only 7500 gallons per hour.

3. Identify Significant Disturbances

The amount of cold water available to cool the reactor can be considered a disturbance because it comes from a reservoir outside of our control. The ambient temperature is also a disturbance. If it drastically increases, the amount of cold water needed to cool the reactor would need to increase as well. Composition of the feed streams will be assumed to be constant in this example. Therefore, they are not considered disturbances.

4. Determine the Type and Location of Sensors

A temperature sensor (TS) and pressure sensor (P) are located on the stream exiting the first reactor vessel. A flow meter (FM) is located on the product stream leaving the second reactor. A composition sensor (CS) will also be located on the product stream leaving the second reactor. The pressure drop can be controlled through the decompressor and thus is a control.

5. Determine the Location of Control Valves

Control valves are located on the feed stream containing the phenol and sodium hydroxide, the incoming cold water to the first heat exchanger, and the sulfuric acid feed stream. There is also a pump located on the carbon dioxide stream that enters the reactor.

6. Perform a Degree of Freedom Analysis

There are 3 valves, 1 pump, and 1decompressor but 5 objectives. This results in zero degrees of freedom. The valve located on the sulfuric acid feed stream is meant to meet the composition constraint placed on the product stream leaving the second reactor. The valve located on the feed stream carrying the reactants is set to satisfy production requirements. The valve on the cold water stream is used to maintain reactor temperature, which satisfies an operational constraint. The pump is to ensure the correct pressure is achieved in the reactor, also satisfying an operational constraint. The decompresser is to maintain a pressure of less than 10atm in the second reactor, thus satisfying another operational constraint.

7. Energy Management

The heat from the exothermic reaction in the first reactor is transferred to the cold water stream. The hot water stream exiting the reactor vessel jacket could be used to heat streams on other processes. The second reactor is assumed to generate negligible heat during the reaction, thus any release of heat from the reactor will be considered safe to release into the environment surrounding the process.

8. Control Process Production Rate and Other Operating Parameters

The production rate is measured by the flow sensor on the product stream and this signals the control valve on the feed stream through a feedback mechanism to change the production rate as necessary.

9. Handle Disturbances and Process Constraints

If the temperature sensor on the reactor exit stream exceeds a certain level due to a diminished cold water supply, the feed stream valve would decrease the amount of reactants entering the reactor. The amount of feed would also be decreased if more than 7500 gallons per hour of cooling water were needed, as this is an operational constraint. If the pressure gauge controlling the pump begins to read higher than allowed pressures, the pump would decrease the flow of the carbon dioxide entering the reactor. Also, if the pressure gauge reads out a pressure that will be too high for the second reactor, the decompresser will be allowed to disperse more pressure. If ambient air temperature drastically increases, the temperature sensor would open the cold water valve allowing more cooling water to enter the reactor vessel jacket. If the composition of the product stream falls below 80 mole percent of salicylic acid, then the valve controlling the sulfuric acid feed would allow more sulfuric acid into the second reactor to increase the conversion of reactants.

10. Monitor Component Balances

The composition sensor and flow meter on the product stream leaving the second reactor will account for every species to ensure that there is no accumulation or loss within the system.

11. Control Individual Unit Operations

The first reactor vessel's pressure is fully controlled by the pressure gauge and pump system and its temperature is fully controlled by the temperature sensor which controls the reactant feed valve and the cool water valve. The second reactor's pressure is fully controlled by the same pressure gauge and the decompresser system, and its temperature will be highly dependent on the amount of cooling water used to cool the product exiting the first reactor.

12. Optimize the Process

Since there are no unaccounted degrees of freedom, there are no valves to adjust in order to optimize the process. It should be noted, however, that if there was no constraint on the composition of the product stream, the sulfuric acid feed valve would have become an unaccounted for degree of freedom. If this had been the case, the valve could be adjusted to maximize the profit of the process. In order to maximize the profit, the benefits of having higher conversion and more product would have to be weighed against the increase costs of using more sulfuric acid feed.

Chapter 2

DYNAMICAL SYSTEMS ANALYSIS

FIXED POINTS

Engineers can gain a better understanding of real world scenarios by using various modeling techniques to explain a system's behaviour. Two of these techniques are ODE modeling and Boolean modeling. An important feature of an accurate ODE model is its fixed point solutions. A fixed point indicates where a steady state condition or equilibrum is reached. After locating these fixed points in a system, the stability of each fixed point can be determined. This stability information enables engineers to ascertain how the system is functioning and its responses to future conditions. It also gives information on how the process should be controlled and helps them to choose the type of control that will work best in achieving this.

Concept Behind Finding Fixed Point

A fixed point is a special system condition where the measured variables or outputs do not change with time. In chemical engineering, we call this a steady state. Fixed points can be either stable or unstable. If disturbances are introduced to a system at steady state, two different results may occur:

1. The system goes back to those original conditions (stable point)
2. The system deviates from those conditions rapidly (unstable point)

ODE Model

When a process or system is modeled by an ODE or a set of ODEs, the fixed points can be found using various mathematical techniques, from basic hand calcuations to advanced mathematical computer programs. Independent of the method used, the basic principle remains the same: The ODE or set of ODEs are set to zero and the independent variables are solved for. At the points where the differential equations equal zero there is no change occurring.

Thus, the solutions found by setting the ODEs equal to zero represent the numerical values of independent variables (*i.e.* temperature, pressure, concentration) at steady state conditions. If a single ODE or set of ODEs becomes too complicated to be solved by hand, a mathematical program such as Mathematica can be used to find fixed points.

Note that in some cases there may not be an analytical method to find a fixed point. This case commonly occurs when the solution to a fixed point involves a high degree polynomial or another mathematical function that does not have an analytical inverse. In these cases, we can still find fixed points numerically if we have the parameters.

Boolean Model

A Boolean Model, as explained in "Boolean Models," consists of a series of variables with two states: True (1) or False (0). A fixed point in a Boolean model is a condition or set of conditions to which the modeled system converges. This is more clearly seen by drawing state transition diagrams.

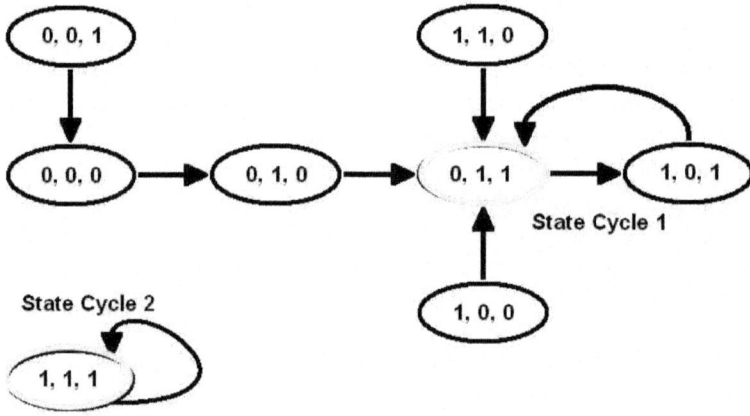

Fig. : State Transition Diagram from Boolean Models.

From the state transition diagram above, we can see that there are two fixed points in this system: 0,1,1 and 1,1,1. Starting in any state on the diagram and following the arrows, one of these two states will be reached eventually, indicating that the system tends to achieve either of these sets of operating conditions. If slight disturbances are introduced to the system while it is operating at one of these sets of conditions, it will return to 0,1,1 or 1,1,1. Also noted in the state transition diagram are state cycles. The difference between a state cycle and a fixed point is that a state cycle refers to the entire set of Boolean functions and transition points leading to the steady-state conditions, whereas a fixed point merely refers to the one point in a state cycle where steady-state conditions are reached (such points are indicated by a yellow circle in the diagram).

Finding Fixed Points: Four Possible Cases

There are four possible scenarios when finding the fixed points of an ODE or system of ODEs:

1. One fixed point
2. Multiple fixed points
3. Infinite fixed points
4. No fixed points

One Fixed Point

The first type of ODE has only one fixed point. An example of such an ODE is found in the Modeling of a Distillation Column. An ODE is used to model the energy balance in the nth stage of the distillation column:

$$\frac{dT_n}{dt} = \frac{1}{M_W}\left[L_{n-1}x_{n-1} - Wx_W\right]\left[T_{n-1} - T_n\right] + \frac{q_r}{M_W c_p}$$

Which can also be written as:

$$\frac{dT_n}{dt} = \frac{1}{M_W}\left[L_{n-1}x_{n-1} - Wx_W\right]\left[T_{n-1}\right] + \frac{q_r}{M_W c_p} + \frac{1}{M_W}\left[L_{n-1}x_{n-1} - Wx_W\right]\left[-T_n\right]$$

If initial conditions *i.e.* $T_{n-1}, L_{n-1}, x_{n-1}$ are known, the equation above reduces to:

$$\frac{dT_n}{dt} = a + bT_n$$

Where a and b are constants since all the variables are now known.

$$a = \frac{1}{M_W}\left[L_{n-1}x_{n-1} - Wx_W\right]\left[T_{n-1}\right] + \frac{q_r}{M_W c_p}$$

$$b = \frac{-1}{M_W}\left[L_{n-1}x_{n-1} - Wx_W\right]$$

By analyzing the equation $\frac{dT_n}{dt} = 0 = a + bT_n$, we can immediately deduce that at steady state $T_n = -\frac{a}{b}$. Clearly, there is only one fixed point in this system, only one temperature of the distillation column which will be at steady-state conditions. We can use Mathematica to solve for the fixed point of this system and check our results. In Mathematica, the Solve[]function can be used to solve complicated equations and systems of complicated equations. There are some simple formatting rules that should be followed while using Mathematica:

1. Type your equation and let the differential be called an arbitrary variable (*e.g.* T[t])
2. Type Solve[T[t]==0,T] and hit Shift+Enter
3. This produces an output contained inside curly brackets

A sample of how the format in Mathematica looks like is shown below:

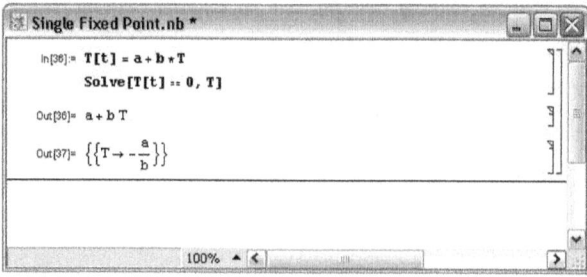

Maple can be used to visualize a single fixed point. Wherever the plot intersects the x-axis represents a fixed point, because the ODE is equal to zero at that point.

The following Maple syntax was used to plot the ODE: plot(0.5+4t, t=-2..2,T=0..5,colour=black);

The constant a = 0.5 and the constant b = 4 in the above example.

The resulting graph is below, the red point indicates at what T a fixed point occurs:

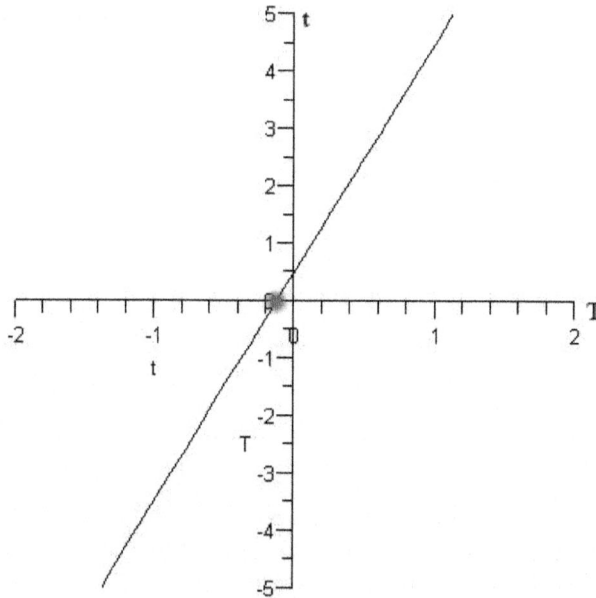

Solving a single fixed point for an ODE and a controller in Mathematica

1. Identify what type of controller it is (P, I, PI, or PID *etc.*
2. Identify your ODE equations (Is the controller a function of the ODE?)

Example: Solve for the fixed points given the three differential equations and the two controllers (u_1 and u_2).

$$\frac{dH}{dt} = (1/A)(F_{(in)} - F_{(out)})$$

$$\frac{dF_{(in)}}{dt} = K_{(v1)}(u_1)$$

$$\frac{dF_{(out)}}{dt} = K_{(v2)}(u_2)H$$

Where H is the level in the tank, Fin is the flow in, Fout the flow out, and u_1 and u_2 are the signals to the valves v_1 and v_2. Kv1 and Kv2 are valve gains (assumed to be linear in this case, although this does not have to be). Note that the exit flow also depends on the depth of fluid in the tank.

You next parameterize your model from experimental data to find values for the constants:

A=2.5 meters squared

K_(v1)=0.046 meters cubed/(minute mA)

K_(v2)=0.017 meters squared/(minute mA)

Next you want to add:

• A full PID controller to regulate Fout via FC1 connected to v2.

• A P-only controller to regulate H via LC1 connected to v1.

For this system you want to maintain the tank level at 3 meters and the exit flow (Fset) at 0.4 m3/minute. The following Mathematica code should look as follows:

```
Finalfixedpts.nb                                                      ─ □ ✕

    u1 = offset1 + Kc1 * (3 - H[t]);
    u2 = offset2 + Kc2 * (0.4 - Fout[t]) + (1 / tauI) * X[t] + tauD * (-Fout'[t]);
    eqns = {X'[t] == 0.4 - Fout[t], H'[t] == (1/A) * (Fin[t] - Fout[t]), Fin'[t] == Kv1 * u1,
        Fout'[t] == Kv2 * u2 * H[t], H[0] == 0, Fin[0] == 0, Fout[0] == 0, X[0] == 0};
    param = {Kv1 → 0.046, Kv2 → 0.017, A → 2.5};
    sol = Solve[eqns /. param /. {H'[t] → 0, Fin'[t] → 0, Fout'[t] → 0, X'[t] → 0},
        {H[t], Fin[t], Fout[t], X[t]}]

    {{Fin[t] → 0.4, X[t] → (-3. Kc1 - 1. offset1) offset2 tauI / (3. Kc1 + offset1), Fout[t] → 0.4, H[t] → 3. Kc1 + offset1 / Kc1}}
```

Multiple Fixed Points

Multiple fixed points for an ODE or system of ODEs indicate that several steady states exist for a process, which is a fairly common situation in reactor kinetics and other applications. When multiple fixed points exist, the optimal steady-state conditions are chosen based on the fixed point's stability and the desired operating conditions of the system.

The following is an example of a system of ODEs with multiple fixed points:

$$\frac{dC_A}{dt} = 14C_A - 2C_A^2 - C_A C_B$$

$$\frac{dC_B}{dt} = 16C_B - 2C_B^2 - C_A C_B$$

The above system of ODEs can be entered into Mathematica with the following syntax:

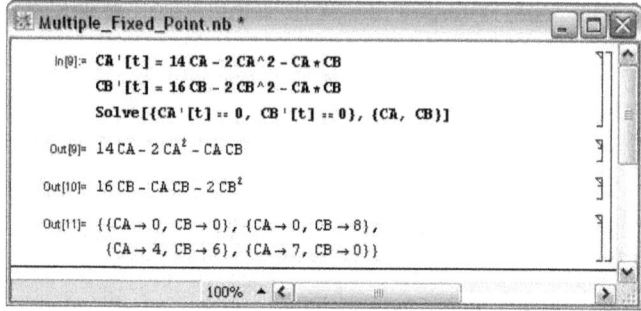

This system in particular has four fixed points. Maple can be used to visualize the fixed points by using the following syntax:

with(plots):

fieldplot([14*x-2*x^2-x*y,16*y-2*y^2-x*y],x=0..10,y=0..10,fieldstrength=log);

The first line initializes the plotting package within Maple that allows for plotting vector fields. The second line uses the command "fieldplot" and inputs the two ODEs that make up the system. The scales of the x and y-axis are set to range from 0 to 10. The fieldstrength command is mainly used for visual purposes, so that the direction of the arrows becomes more apparent. Below is the resulting plot:

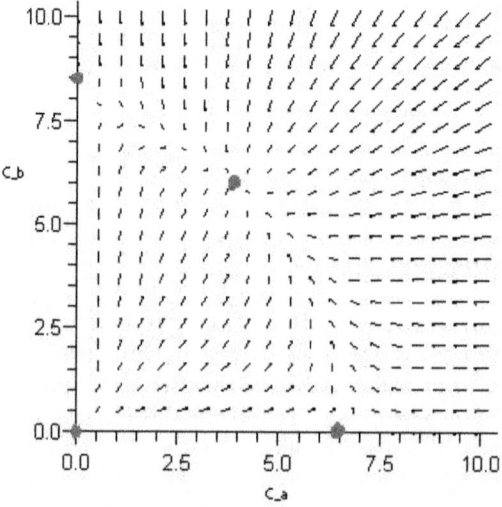

The red dots indicate the fixed points of the system. On the plot, these points are where all the surrounding arrows converge or diverge. Converging arrows indicate a stable fixed point, in this example the point at (4,6) is a stable fixed point. Diverging arrows indicate an unstable fixed point, in this example (0,0), (0,8) and (7,0) are unstable fixed points.

Infinite Fixed Points

An example of an ODE with infinite fixed points is an oscillating ODE such as:

$$\frac{dy}{dx} = cos(ax)$$

where a is a constant.

Using Mathematica to solve for the fixed points by setting

$$\frac{dy}{dx} = 0 = cos(ax)$$

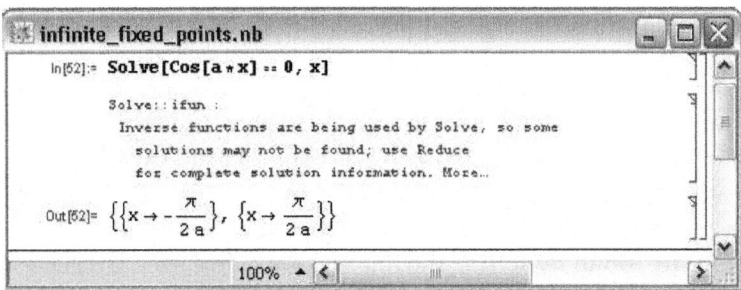

If you click the "More" link on Mathematica it will basically state that there are other solutions possible according to the Help section shown below:

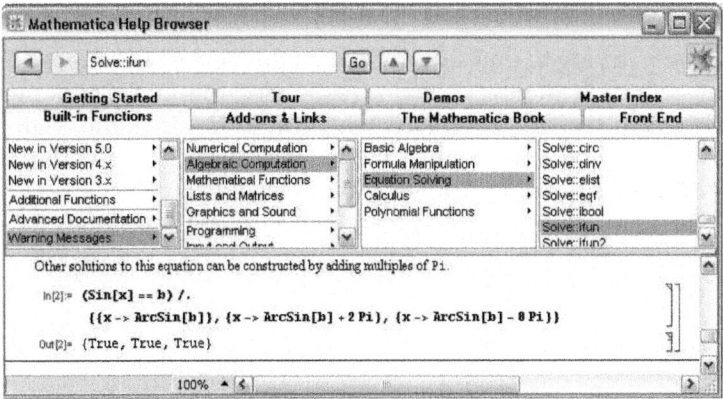

The Maple syntax used to graph the solved differential equation is:

plot(cos(3t),t=0..10,T=-1..1,colour=black);

The constant a = 3 in this case.

The infinite fixed points can be seen in the graph below, where anytime the function crosses the x axis, we have a fixed point:

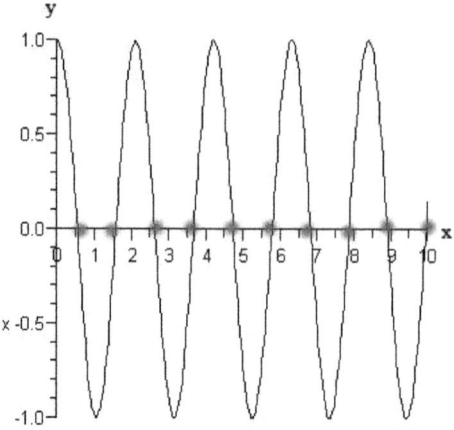

No Fixed Points

The fourth type of ODE does not contain fixed points. This occurs when a certain variable (such as temperature or pressure) has no effect on a system regardless of how it changes. Generally, systems with this sort of behaviour should be avoided because they are difficult to control as they are always changing.

This can be modeled by vertical or horizontal lines due to the fact that no fixed points are found by setting the line equal to zero. An ODE is used to model a line held constant at a:

$$\frac{dT}{dt} = a$$

Where, a can be any constant except 0.

Intuitively, trying to find a fixed point in this system is not possible, because a constant such as 3 can never equal zero. Solving this ODE is not possible even by analyzing the system. Therefore, when inputting this into Mathematica, it yields {}. The notation {} means that there are no fixed points within the system. The image below is how Mathematica solves the ODE.

By using Maple (version 10), one can visually see a lack of fixed points by using the following syntax:

plot(3, t = 0..10, T = 0..10, colour = black);

The constant a = 3 in the above case.

This image shows that the line is horizontal and never crosses the x axis, indicating a lack of fixed points.

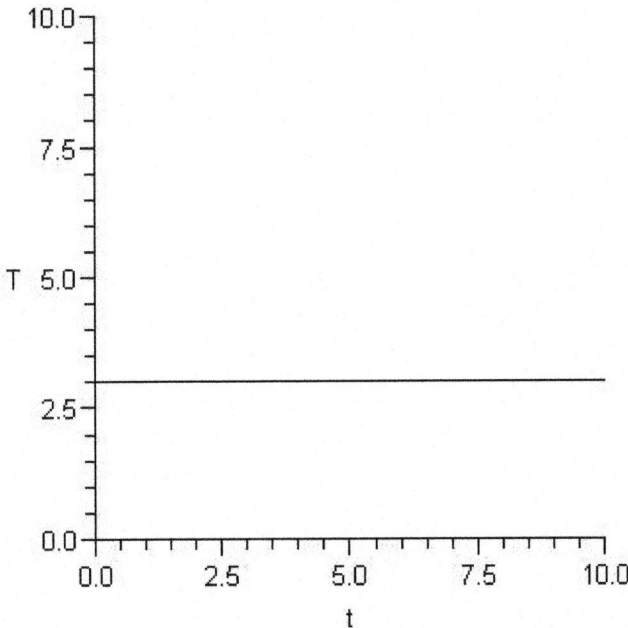

Linearizing ODEs

Chemical engineering processes often operate in nonlinear and unsteady manners (*i.e.* not always at steady state), and are generally governed by nonlinear ordinary differential equations (ODEs). The ODE is a relation that contains functions of only one independent variable and derivatives with respect to that variable. Many studies have been devoted to developing solutions to these equations, and in cases where the ODE is linear it can be solved easily using an analytical method. However, if the ODE is nonlinear and not all of the operating parameters are available, it is frequently difficult or impossible to solve equations directly. Even when all the parameters are known, powerful computational and mathematical tools are needed to completely solve the ODEs in order to model the process. In order to simplify this modeling procedure and obtain approximate functions to describe the process, engineers often linearize the ODEs and employ matrix math to solve the linearized equations.

A linear equation is an equation in which each term is either a constant or the product of a constant times the first power of a variable. These equations are called "linear" because they represent straight lines in Cartesian coordinates. A common form of a linear equation in the two variables x and y is $y = mx + b$. This is opposed to a nonlinear equation, such as $m = e^x + x^2 + 2x + 5$. Even though $2x + 5$ is a linear portion of the equation, e^x and x^2 are not. Any nonlinear terms in an equation makes the whole system nonlinear.

Non-linear system of equations:

$$\frac{dA}{dt} = 3A^2 + 2B + C - 7D^3$$

$$\frac{dB}{dt} = A + C^2 + 2D$$

$$\frac{dC}{dt} = A + 4B^2 - C^2$$

$$\frac{dD}{dt} = 2C - D$$

After linearization (around the steady state point {-0.47,-0.35,0.11,0.23}:

$$\begin{pmatrix} A' \\ B' \\ C' \\ D' \end{pmatrix} = \begin{pmatrix} -2.83 & 2 & 1 & 0 \\ 1 & 0 & 0.23 & 0 \\ 1 & -2.78 & -0.23 & 0 \\ 0 & 0 & 2 & 0 \end{pmatrix} \begin{pmatrix} A \\ B \\ C \\ D \end{pmatrix} + \begin{pmatrix} k_1 \\ k_2 \\ k_3 \\ k_4 \end{pmatrix}$$

Note that each equation is comprised solely of first order variables.

Even though it is unlikely that the chemical engineering process to be modeled operates in a linear manner, all systems can be approximated as linear at a point. This is preferred as linear systems are much easier to work with than nonlinear equations. Although linearization is not an exact solution to ODEs, it does allow engineers to observe the behaviour of a process. For example, linearized ODEs are often used to indicate exactly how far from steady state a given process deviates over specified operating ranges.

Applications to Chemical Engineering

As mentioned above, linearizing ODEs allows engineers to understand the behaviour of their system at a given point. This is very important because many ODEs are impossible to solve analytically. It will also lead to determining the local stability of that point. Most of the time a system will be linearized around steady state, but this is not always the case. You may be interested in understanding the behaviour of your system at its operating point or equilibrium state (not necessarily steady state). The linearization approach can be used for any type of nonlinear system; however, as a chemical engineer, linearizing will usually involve ODEs. Chemical engineers use ODEs in applications such as CSTRs, heat exchangers, or biological cell growth.

It is also important to understand the advantages and disadvantages of linearizing a system of ODEs:

Advantages

- Provides a simpler, more convenient way to solve the ODEs
- The behaviour of a process can be observed
- Any type or order of ODE can be used

Disadvantages

- The solution is only an exact solution at the chosen point; otherwise it is an approximation and becomes less accurate away from the point
- Although linearizing is a quicker alternative, it takes time to initially learn the process (ex: using Mathematica)

Another use for linearization of the equations that govern chemical processes is to determine the stability and characteristics of the steady states.

General Procedure for Linearization

Linearization is the process in which a nonlinear system is converted into a simpler linear system. This is performed due to the fact that linear systems are typically easier to work with than nonlinear systems. For this course, the linearization process can be performed using Mathematica. The specific instructions on how to do this can be found below.

1. Choose a relevant point for linear approximation, two options available are:
   - Steady state- points where system does not change
   - Current location- given where you are now
2. Calculate the Jacobian matrix at that point. The Jacobian is essentially a Taylor series expansion.
3. Solve to find unknown constants using algebraic methods.

Linearization by Hand

In order to linearize an ordinary differential equation (ODE), the following procedure can be employed. A simple differential equation is used to demonstrate how to implement this procedure, but it should be noted that any type or order of ODE can be linearized using this procedure.

1. Use a Taylor series expansion (truncating after the linear terms) to approximate the right-hand side of the ODE.

   Let's say we start with the following ODE: $\dfrac{dx}{dt} = f(x) = 3x^2$. This ODE de-

scribes the behaviour of some variable, x, with respect to time.

A Taylor series is a series expansion of a function about a point. If x= a, an expansion of a real function is given by:

$$f(x) = \sum_{n=0}^{\infty} \frac{f^n(a)}{n!}(x-a)^n$$

$$f(x) = f(a) + f'(a)(x-a) + \frac{f^2(a)}{2!}(x-a)^2 + \frac{f^3(a)}{3!}(x-a)^3 + ... + \frac{f^n(a)}{n!}(x-a)^n + ...$$

When x=0, the function is also known as Maclaurin series. Taylor's theorem states that any function satisfying certain conditions can be expressed as a Taylor series.

For simplicity's sake, only the first two terms (the zero- and first-order) terms of this series are used in Taylor approximations for linearizing ODEs. Additionally, this truncation (*i.e.* "chopping" off the n=2 and higher terms from the polynomial shown above) assures that the Taylor Series is a linear polynomial. If more terms are used, the polynomial would have $(x - a)^2$ and higher order terms and become a nonlinear equation. The variable 'a' in the Taylor series is the point chosen to linearize the function around. Because it is desired that most processes run at steady state, this point will be the steady state point. So, our differential equation can be approximated as:

$$\frac{dx}{dt} = f(x) \approx f(a) + f'(a)(x - a) = f(a) + 6a(x - a)$$

Since a is our steady state point, f(a) should always be equal to zero, and this simplifies our expression further down to:

$$\frac{dx}{dt} = f(x) \approx f'(a)(x - a) = 6a(x - a)$$

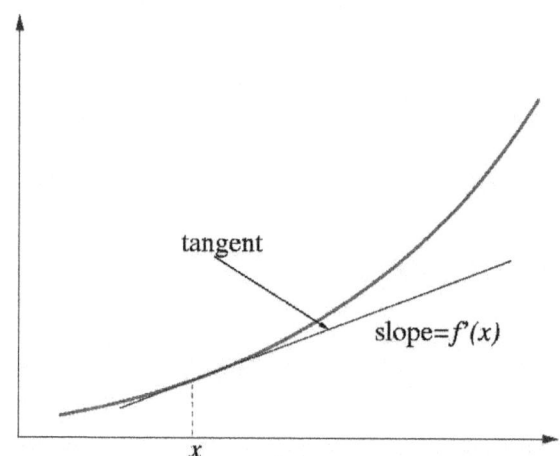

The graph shown above shows the approximation of f(x) at (x,f(x)). As mentioned previously, linearization is only an approximation for any given function near a continuous point. When working with a system of ODEs, the Jacobian is written as a matrix. It is the matrix of constants needed to describe a system's linearity. The Jacobian may be thought of as how much a system is distorted to take on a linear identity. A jacobian matrix will always be a square(#rows = #columns) and it shows how each equation varies with each variable. The Jacobian matrix is defined as:

$$J(x_1, \ldots, x_n) = \begin{bmatrix} \dfrac{\partial y_1}{\partial x_1} & \cdots & \dfrac{\partial y_1}{\partial x_n} \\ \vdots & \ddots & \vdots \\ \dfrac{\partial y_n}{\partial x_1} & \cdots & \dfrac{\partial y_n}{\partial x_n} \end{bmatrix}.$$

And is used as such:

Nonlinear system

$$\frac{dx}{dt} = 2x^3 + \cos(5y) - 12\arctan(z-3)$$

$$\frac{dy}{dt} = 7y^3 + \sin(5x)$$

$$\frac{dz}{dt} = 1.3\log(x+y)$$

Linearized system

$$\begin{bmatrix} x' \\ y' \\ z' \end{bmatrix} = \begin{bmatrix} J_1 & J_2 & J_3 \\ J_4 & J_5 & J_6 \\ J_7 & J_8 & J_9 \end{bmatrix} \begin{bmatrix} x \\ y \\ z \end{bmatrix} + \begin{bmatrix} c_1 \\ c_2 \\ c_3 \end{bmatrix}$$

Jacobian matrix

Example

Lets say you have the following set of equations and you want to find its jacobian matrix around the point A=3,B=2.

$$\frac{dA}{dt} = 3A - A^2 - AB$$

$$\frac{dB}{dt} = 6B - AB - 2B^2$$

We find the jacobian by taking the derivative of each equation with respect to each variable.

$$\frac{d(3A - A^2 - AB)}{dA} = 3 - 2A - B, \frac{d(3A - A^2 - AB)}{dB} = -A$$

$$\frac{d(6B - AB - 2B^2)}{dA} = -B, \frac{d(6B - AB - 2B^2)}{dB} = 6 - A - 4B$$

These are the equations in the matrix. The values of the variables from whatever point we are linearizing are then put into these equations and calculated out to get the Jacobian.

$$Jac = \begin{bmatrix} 3 - 2A - B & -A \\ -B & 6 - A - 4B \end{bmatrix} A = 4, B = 2$$

$$Jac = \begin{bmatrix} -7 & -4 \\ -2 & -6 \end{bmatrix}$$

2. Change the approximation by linearizing around a steady state point in order to describe how the process deviates from steady state.

The following substitution can be made:

$$\frac{dx}{dt} = \frac{d(x-a)}{dt}$$

$$\frac{d(x-a)}{dt} \approx 6a(x-a)$$

This substitution is allowed because 'a' is a constant, and the derivative of a constant is zero.

Substituting (x-a) for x signifies that our differential equation now shows how our function, x, deviates away from the steady state value, a, with respect to time. This deviation, (x-a), is commonly expressed as x'. It should also be noted that the quantity '6a' is a constant, and thus will be further recognized as 'A'.

Our final linearized equation becomes:

$$\frac{dx'}{dt} \approx Ax'$$

The once nonlinear ODE, $\frac{dx}{dt} = f(x) = 3x^2$, has now been simplified into a linear differential equation.

The procedure of linearization typically occurs around the steady state point or points of a specified process. Engineers anticipate a certain change in output for the particular steady state point, and may proceed to linearize around it to complete their approximation. Please follow the graphic below for further detail. Note that a steay state point occurs when $dx/dt = 0$

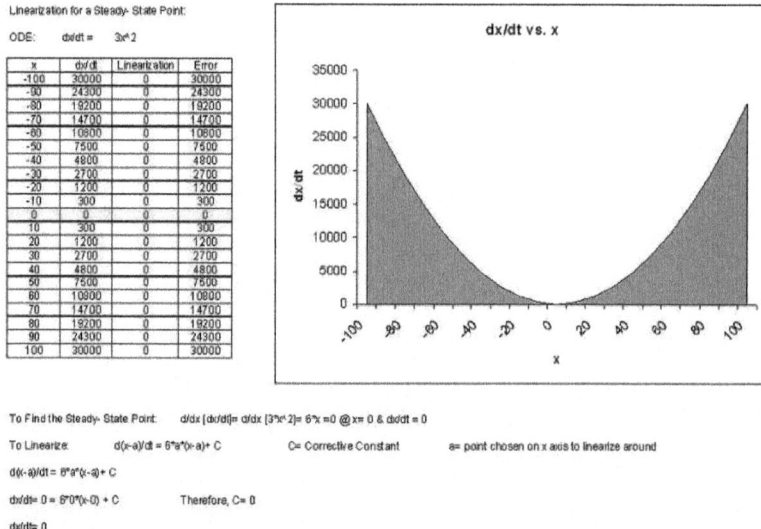

As can be seen, moving farther away from the steady state point results in significantly larger deviation, and thus error (actual-linearization).

Occasionally, for very unique operating conditions, plant management may decide to momentarily run a process outside of strict steady state conditions (perhaps a unique start-up procedure, shut-down recovery, *etc.*. To reinforce the concept of linearization around an unsteady state point (arbitrarily chosen in this example), please consider the following visual representation. Note that a=50 and C=7500 in the first plot.

Differential Equation: dx/dt= f(x) = 3*x^2   C is held constant at 7,500 for the first plot.

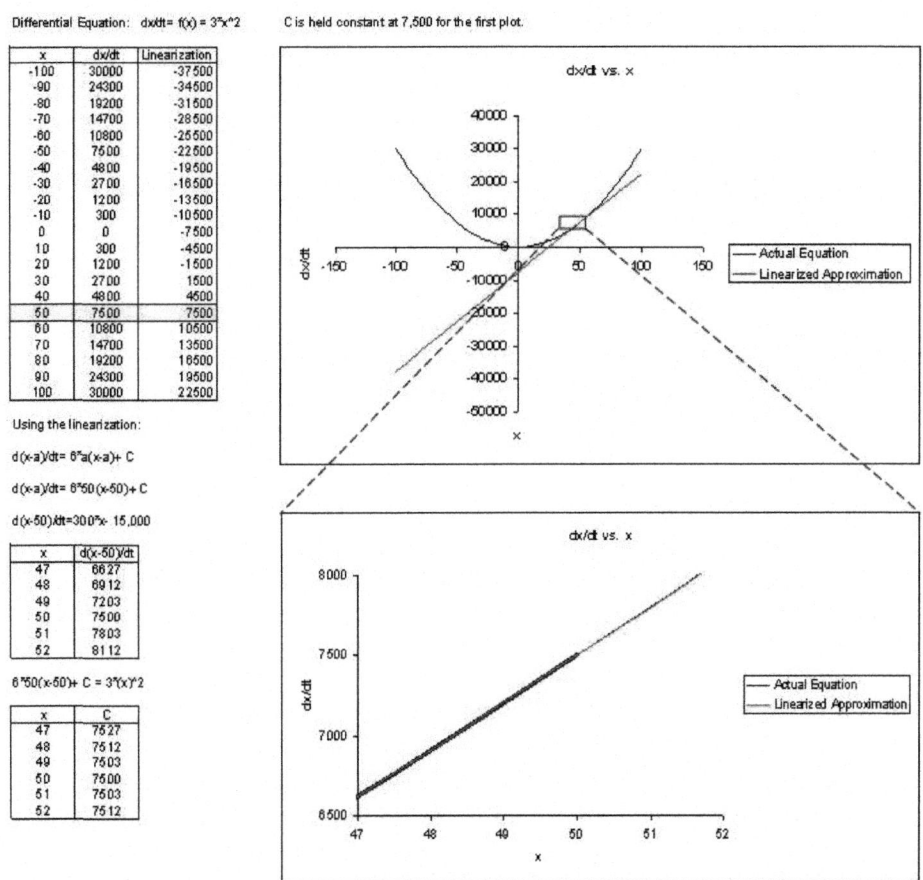

x	dx/dt	Linearization
-100	30000	-37500
-90	24300	-34500
-80	19200	-31500
-70	14700	-28500
-60	10800	-25500
-50	7500	-22500
-40	4800	-19500
-30	2700	-16500
-20	1200	-13500
-10	300	-10500
0	0	-7500
10	300	-4500
20	1200	-1500
30	2700	1500
40	4800	4500
50	7500	7500
60	10800	10500
70	14700	13500
80	19200	16500
90	24300	19500
100	30000	22500

Using the linearization:

$d(x-a)/dt = 6*a(x-a) + C$

$d(x-a)/dt = 6*50(x-50) + C$

$d(x-50)/dt = 300*x - 15,000$

x	d(x-50)/dt
47	6627
48	6912
49	7203
50	7500
51	7803
52	8112

$6*50(x-50) + C = 3*(x)^2$

x	C
47	7527
48	7512
49	7503
50	7500
51	7503
52	7512

The second plot is a magnification of a small section of the first plot. As one can readily notice, both the linear approximation and the actual graph overlap almost exactly over this small range. This illustrates how this particular linearization could be used to approximate this region of the function and can describe its behaviour quite accurately. This approximating technique using the linearization of ODE can be performed around different points (different values of a) using the same method in order to model the behaviour of the rest of the dx/dtvs x function. Note: frequent recalculation of the integration constant, C, permits for increased accuracy in approximation.

Example of a Simple Linearization Process in Use

Provided a circumstance in which developing a rigorous analytical solution is unfeasible due to time, access to computing resources, or mathematical ability, the linearization process offers a convenient and swift alternative.

An appropriate introduction to linearization application is featured in section 5.3 of the Bequette reading and has been described below.

The following model describes the behaviour of tank height as a function of time (t), faced with a steady- state incoming liquid flowrate. dh/dt represents the change in height of the fluid contained within the tank as a function of time, h indicates the height (or level) or the fluid contained within the tank, F represents the magnitude of incoming flow. The following model describes the tank level/ height fluctuation as function of both the current height (h) and flow (F). Fluid enters the tank as flow per area of the inlet(F/A), and leaves as a function of a corrective proportional term, β, also per area of the outlet, multiplied by the square root of the current height.

Instead of fluid leaving through this term, we may also consider it a form of feedback control, which subsequently reduces the amount of inlet fluid instead (depends on the application, the final result is identical). Therefore:

$$\frac{dh}{dt} = f(h, F) = \frac{F}{A} - \frac{\beta}{A} * \sqrt{h}$$

Important note about the relevance of the provided variables:

F_s indicates the initial flow rate for the described system and is important in calculating the intercept and subsequently the calculative corrective term for the linearization (if the linearization is not performed at steady-state, as in this example). h_s describes the tank height for the specified flow rate, F_s. Upon plugging in the value of h_s into the linearized formula, you will notice that $f(h,F_s) = 0$, the point on the plot where both the linearization and the characteristic equation coincide (*i.e.* are tangent).

h remains a variable for the derivation so that students may notice that it is a dependant variable (dependent upon the flowrate). Bequette must (and does) provide both the values for F_s, h_s so that a linearization may proceed (otherwise we will simply have an equation describing the slope with no characteristic point to pass through).

System Parameters (F_s is an example flow rate at steady state, this information is used to determine the intercept of the linearized formula):

$$A = 1ft^2; h_s = 5ft; \beta = \frac{1}{\sqrt{5}}\frac{ft^{2.5}}{min}; F_s = \frac{1ft^3}{min}$$

Rewriting the system with the input variables:

$$\frac{dh}{dt} = f(h, F_s) = 1 + \frac{1}{\sqrt{5}} * \sqrt{h}$$

To linearize around a certain point, simply evaluate the derivative of the desired function and add in a corrective constant, C, represented by the value of the function at the initial (specified) condition.

$$f(h, F_s)approx. = f(h, F_s) + \frac{\partial f}{\partial h}|_{h, F_s}(h - h_s)$$

$$f(h, F_s)approx. = 0 + \frac{-1}{2\sqrt{5}\sqrt{h_s}}|_{h, F_s}(h - h_s)$$

Selecting, h=5 to linearize around, we present an adapted version of the text example (linearization is in red):

$$f(h, 1)approx. = 0 + \frac{-1}{2\sqrt{5}\sqrt{5}}|_{h,F_s}(h-5)$$

$$f(5, 1)approx. = 0 + \frac{-1}{2\sqrt{5}\sqrt{5}}|_{h,F_s}(5-5) = 0$$

$$f(h,1)approx. = -0.1 * (h-5)$$

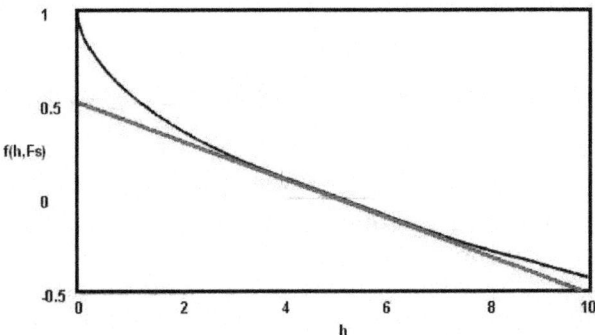

The physical significance of this example:

The model describes a situation where a tank maintains the steady state set point, a level (height) of 5 feet, by regulating the input (and/or output) flow rate(s). A height above the preferred level of 5 ft. results in a negative height change per time; conversely, a height below the preferred level of 5 ft. results in a positive height change per time, so as to always approach the steady state level.

Also, please note that the linearized system responds relatively slower than the characteristic equation system (non- linear)when tank level falls below 5 ft. and faster than the characteristic equation system when the tank level is above 5 ft.

Please read on below to learn a step-by-step technique on how to complete this method for most equations.

Linearization using Mathematica

Most ODEs are not as simple as the example worked out above. For those that can be solved analytically, the 'by-hand' method from the previous section can be used, although the math becomes tricky and tedious when one is dealing with systems of ODEs. For those that cannot be solved analytically and can only be approximated numerically, it is sometimes impossible to do the numerical analysis by hand. For these ODEs, powerful computational tools such as Mathematica can be used to perform linearization.

The following Mathematica file contains a worked out example (CSTR w/ heat exchange). This example contains multiple ODEs and multiple types of variables. When dealing with ODEs, there are three main type of variables: state, input, and output. State variables describes the system at any given time (*i.e.,*Ca, Cb, T, *etc.*. Input variables are simply inputs into the system. Output variables are those that are not directly related to the inputs, but are dependent on the state variables (*i.e.,* k, the reaction constant). You will discover in the CSTR example that it is necessary to linearize state variables and output variables separately.

Before proceeding to the Mathematica file we will first present some of the Mathematica commands that will be necessary to linearize the CSTR example.

Note that you need to press shift+enter(shift+return on Mac) after a command to evaluate the expression.

Example:

In[1] = ((0.04587*54.32497+4.59058433734947^7)/48700043)^131.94564

Out[1] = 0.000411241

Another thing to note is the use of assignment and equality operators.

= is called **immediate assignment operator**and := is called **delayed assignment operator**. When immediate assignment operator is used, the right hand side of the operator is evaluated every time an assignment is made. However, when delayed assignment operator is used, the right hand side is evaluated only when the value of the expression on the left hand side is requested.

lhs=rhs   rhs is intended to be the "final value" of lhs (*e.g.,* f[x_]=1-x^2)

lhs:=rhs   rhs gives a "command" or "program" to be executed whenever you ask for

the value of lhs (*e.g.,* f[x_]:=Expand[1-x^2])

p := x^3 - 6 x^2 + 11 x - 6

This input, which produces no output, assigns the polynomial expression to the variable p. The assignment is done because this polynomial may be used again later, so the assignment saves retyping it. The **equality operator ==**, which is used here to form the equation we are asking Mathematica to solve:

In[2]:= Solve[p == 0]

Out[2]= {{x ->1}, {x ->2}, {x ->3}}

When solving nonlinear ODEs using Mathematica, it is necessary to form a matrix and there are several commands that can be used to create matrices.

D[X,Y]

This command takes the partial derivative of an expression (X) with respect to a variable (Y) defined by the user. The first term in the square brackets is the experession you wish to differentiate, the second is the variable you wish to differentiate with respect to.

AppendRows[column1,column2]

This command creates a matrix by combining two columns. In order to use this command you must first load the Matrix Manipulation package using the command <<LinearAlgebra`MatrixManipulation`

M = {{x1,x2,x3},{y1,y2,y3},{z1,z2,z3}}

This command creates a matrix, in this case a 3x3 matrix.

MatrixForm[X]

This command displays a specified matrix in matrix form.

These commands will allow us to proceed with the linearization of the CSTR problem shown below.

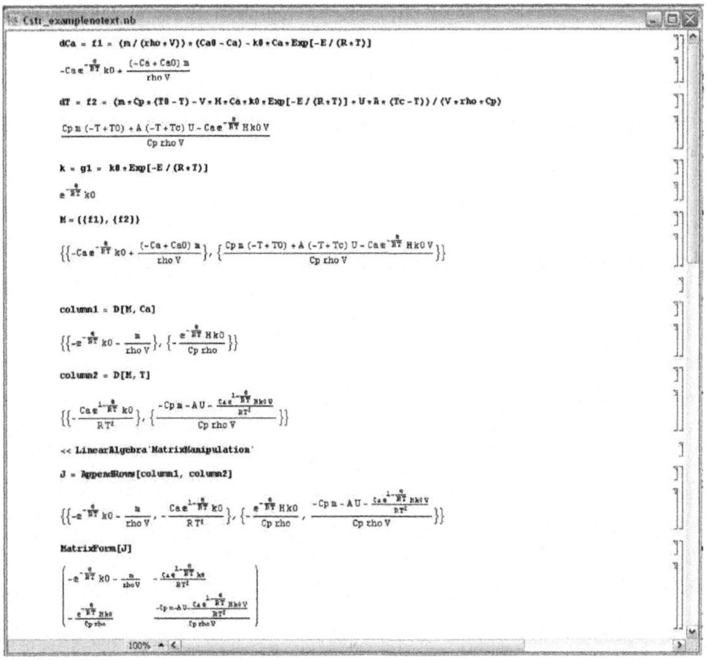

The first three commands input the equations that govern the behaviour of the concentration, Ca, the temperature, T, and the rate of reaction, k. Once the equations have been entered into Mathematica a matrix, M, with one column and two rows containing the equations for Ca and T is created. Next two new matrices are created by taking the derivative of matrix M is with respect to both Ca and T. These matrices, called column1 and column2 respectively, together form the Jacobian of the system and is then displayed in matrix form. The **Jacobian matrix** is the matrix of all first-order partial derivatives of a vector-valued function. Its importance lies in the fact that it represents the best linear approximation to a differentiable function near a given point. For example, given a set y=f(x) of n equations in variables $x_1, x_2,...,x_n$, the Jacobian matrix is defined as following:

$$ J(x_1, \ldots, x_n) = \begin{bmatrix} \dfrac{\partial y_1}{\partial x_1} & \cdots & \dfrac{\partial y_1}{\partial x_n} \\ \vdots & \ddots & \vdots \\ \dfrac{\partial y_n}{\partial x_1} & \cdots & \dfrac{\partial y_n}{\partial x_n} \end{bmatrix}. $$

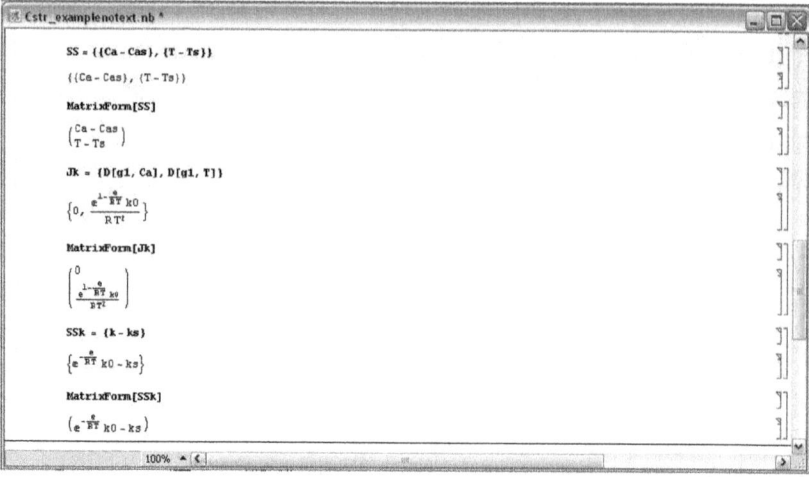

Now that we have the Jacobian we need to create a matrix containing the deviation from steady state for each of the variables. This matrix, SS, contains the actual concentration and temperature, C and T, minus the steady state concentration and temperature, Cas and Ts. The matrix is then displayed in matrix form. We now have both the Jacobian and the deviation matrix for the state variables. The next four commands create the Jacobian and deviation matrix for the output variable, k. The first command creates the Jacobian matrix by taking the derivative of the k equation with respect to Ca and T. The Jacobian is then shown in matrix form. Finally the deviation matrix for k is created in the same manner as above and then displayed in matrix form. Note that because k is defined above, this expression is substituted in for k in the deviation matrix. The following Mathematica file contains the code shown above with extra comments explaining why each step is performed Media:cstr_example.nb. It may be useful to downolad this file and run the program in Mathematica yourself to get a feel for the syntax. Downloading the file will also allow you to make any changes and edits to customize this example to another example of interest.

Note: This file needs to be saved to your computer, and then opened using Mathematica to properly run.

To see another example lets linearize the first 4 differential equations given in the introduction section.

Non-linear system of equations:

$$\frac{dA}{dt} = 3A^2 + 2B + C - 7D^3$$

$$\frac{dC}{dt} = A + 4B^2 - C^2$$

$$\frac{dD}{dt} = 2C - D$$

After linearization (around the steady state point {-0.47,-0.35,0.11,0.23}:

$$\begin{pmatrix} A' \\ B' \\ C' \\ D' \end{pmatrix} = \begin{pmatrix} -2.83 & 2 & 1 & -1.10 \\ 1 & 0 & 0.23 & 2 \\ 1 & -2.78 & -0.23 & 0 \\ 0 & 0 & 2 & -1 \end{pmatrix} \begin{pmatrix} A \\ B \\ C \\ D \end{pmatrix} + \begin{pmatrix} 0.50 \\ 0.01 \\ 0.47 \\ 0 \end{pmatrix}$$

```
lineODE.nb *

In[10]:=  eqns = {3*a^2 + 2*b + c - 7*d^3, a + c^2 + 2*d, a + 4*b^2 - c^2, 2*c - d};
          (*solving for fixed points under s1*)
          s1 = NSolve[{eqns[[1]] == 0, eqns[[2]] == 0, eqns[[3]] == 0, eqns[[4]] == 0}, {a, b, c, d}]

Out[11]=  {{a → -113.663, b → 6.92708, c → 8.84727, d → 17.6945}, {a → -114.027, b → -6.93898, c → 8.86402, d → 17.728},
          {a → -11.0743, b → 1.9117, c → 1.88257, d → 3.76514}, {a → -10.27, b → -1.83233, c → 1.77756, d → 3.55512},
          {a → 0.308224 - 0.364633 i, b → 0.140184 + 0.312248 i, c → -0.0762646 + 0.0947721 i, d → -0.152529 + 0.189544 i},
          {a → 0.308224 - 0.364633 i, b → 0.140184 - 0.312248 i, c → -0.0762646 - 0.0947721 i, d → -0.152529 - 0.189544 i},
          {a → -0.470879, b → -0.347842, c → 0.114445, d → 0.228891}, {a → 0., b → 0., c → 0., d → 0.}}

In[12]:=  (*s1[[7]] equals {a→-0.47, b→-0.35, c→0.11, d→0.23} which is the steady state then define the Jacobian at the s1[[7]]*)
          Jac = {{D[eqns[[1]], a], D[eqns[[1]], b], D[eqns[[1]], c], D[eqns[[1]], d]},
                 {D[eqns[[2]], a], D[eqns[[2]], b], D[eqns[[2]], c], D[eqns[[2]], d]},
                 {D[eqns[[3]], a], D[eqns[[3]], b], D[eqns[[3]], c], D[eqns[[3]], d]},
                 {D[eqns[[4]], a], D[eqns[[4]], b], D[eqns[[4]], c], D[eqns[[4]], d]}} /. s1[[7]];
          (*The Jacobian is changed into Matrix Form*)
          MatrixForm[Jac]

Out[13]//MatrixForm=
          ( -2.82528  2         1          -1.10021 )
          (  1        0         0.228891   2        )
          (  1       -2.78274  -0.228891   0        )
          (  0        0         2         -1        )

In[14]:=  (*To calculate the constants aka k1 k2 k3 and k4 Define m1 as 2nd matrix on righthand side of linearization e/m*)
          m1 = {a, b, c, d}
          (*Execute the dot product of m1 and the Jacobian matrix*)
          Jac.m1

Out[14]=  {a, b, c, d}

Out[15]=  {-2.82528 a + 2 b + c - 1.10021 d, a + 0.228891 c + 2 d, a - 2.78274 b - 0.228891 c, 2 c - d}

In[16]:=  (*Determine the dot product of m1 at the steady state values*)
          Jac.m1 /. s1[[7]]

Out[16]=  {0.497297, 0.0130977, 0.470879, 0.}
```

The 4 differential equations above are added into a Mathematica code as "eqns" and "s1" is the fixed points of the differentials. The steady state values found for "a, b, c, and d" are called "s1doubleBrackets(7)" After the steady state values are found, the Jacobian matrix can be found at those values.

To find "k1, k2, k3, and k4" the constants of the Linearization matrix equation, "m1" must be defined, which is the 2nd matrix on the right-hand side of the Linearization matrix equation.

To determine the k values (in matrix form), execute the dot product of "m1" and the "Jac" matrix, which is done by the "." operator. Therefore it should look like "Jac.m1"

To obtain the k values, determing the "Jac.m1" at the steady state values, which is done by the "/." operator. Therefore it should look like "Jac.m1/.s1doubleBrackets(7)"

EIGENVALUES AND EIGENVECTORS

Eigenvectors(v) and Eigenvalues(λ) are mathematical tools used in a wide-range of applications. They are used to solve differential equations, harmonics problems, population models, *etc.* In Chemical Engineering they are mostly used to solve differential equations and to analyze the stability of a system.

Defintion of Eigenvector and Eigenvalues:

An **Eigenvector** is a vector that maintains its direction after undergoing a linear transformation.

An **Eigenvalue** is the scalar value that the eigenvector was multiplied by during the linear transformation.

Eigenvectors and Eigenvalues are best explained using an example. Take a look at the picture below.

In the left picture, two vectors were drawn on the Mona Lisa. The picture then under went a linear transformation and is shown on the right. The red vector maintained its direction; therefore, it's an eigenvector for that linear transformation. The blue vector did not maintain its director during the transformation; thus, it is not an eigenvector. The eigenvalue for the red vector in this example is 1 because the arrow was not lengthened or shortened during the transformation. If the red vector, on the right, were twice the size than the original vector then the eigenvalue would be 2. If the red vector were pointing directly down and remained the size in the picture, the eigenvalue would be -1.

Now that you have an idea of what an eigenvector and eigenvalue are we can start talking about the mathematics behind them.

Fundamental Equation

The following equation must hold true for Eigenvectors and Eigenvalues given a square matrix **A**:

$$\mathbf{A} \cdot \mathbf{v} = \lambda \cdot \mathbf{v}$$

Where:

A is a square matrix

v is the Eigenvector

λ is the Eigenvalue

Let's go through a simple example so you understand the fundamental equation better.

Question:

Is $v = \begin{bmatrix} 1 \\ -2 \end{bmatrix}$ an eigenvector with the corresponding $\lambda = 0$ for the matrix

$A = \begin{bmatrix} 6 & 3 \\ -2 & -1 \end{bmatrix}$?

Answer:

$$A \cdot v \quad = \quad \lambda \cdot v$$

$$\begin{bmatrix} 6 & 3 \\ -2 & -1 \end{bmatrix} \cdot \begin{bmatrix} 1 \\ -2 \end{bmatrix} = 0 \begin{bmatrix} 1 \\ -2 \end{bmatrix}$$

$$\begin{bmatrix} 0 \\ 0 \end{bmatrix} = \begin{bmatrix} 0 \\ 0 \end{bmatrix}$$

Therefore, it is true that v and λ are an eigenvector and eigenvalue respectively, for **A**.

Calculating Eigenvalues and Eigenvectors

Calculation of the eigenvalues and the corresponding eigenvectors is completed using several principles of linear algebra. This can be done by hand, or for more complex situations a multitude of software packages (*i.e.*,Mathematica) can be used. The following discussion will work for any $n\text{x}n$ matrix; however for the sake of simplicity, smaller and more manageable matrices are used.

LINEAR ALGEBRA REVIEW

For those who are unfamiliar with linear algebra, this section is designed to give the necessary knowledge used to compute the eigenvalues and eigenvectors. For a more extensive discussion on linear algebra, please consult the references.

Basic Matrix Operations

An $m \text{ x } n$**matrix A** is a rectangular array of mn numbers (or elements) arranged in horizontal **rows**(m) and vertical **columns**(n):

$$A = \begin{bmatrix} a_{11} & a_{1j} & a_{1n} \\ a_{i1} & a_{ij} & a_{in} \\ a_{m1} & a_{mj} & a_{mn} \end{bmatrix}$$

To represent a matrix with the element aij in the ith row and jth column, we use the abbreviation $A = [aij]$. Two $m \text{ x } n$ matrices $\mathbf{A} = [aij]$ and $\mathbf{B} = [bij]$ are said to be equal if corresponding elements are equal.

Addition and Subtraction

We can add **A** and **B** by adding corresponding elements: $\mathbf{A}+\mathbf{B}=[aij]+[bij]=[aij+bij]$ This will give the element in row i and column j of $\mathbf{C}=\mathbf{A}+\mathbf{B}$ to have $cij = aij + bij$.

More detailed addition and subtraction of matrices can be found in the example below.

$$\begin{bmatrix} 1 & 2 & 6 \\ 4 & 5 & 10 \\ 5 & 3 & 11 \end{bmatrix} + \begin{bmatrix} 8 & 3 & 5 \\ 5 & 4 & 4 \\ 3 & 0 & 6 \end{bmatrix} = \begin{bmatrix} 1+8 & 2+3 & 6+5 \\ 4+5 & 5+4 & 10+4 \\ 5+3 & 3+0 & 11+6 \end{bmatrix} = \begin{bmatrix} 9 & 5 & 11 \\ 9 & 9 & 14 \\ 8 & 3 & 17 \end{bmatrix}$$

Multiplication

Multiplication of matrices are NOT done in the same manner as addition and subtraction. Let's look at the following matrix multiplication: $\mathbf{A} * \mathbf{B} = \mathbf{C}$ **A** is an $m \times n$ matrix, **B** is an $n \times p$ matrix, and **C** is an $m \times p$ matrix. Therefore the resulting matrix, **C**, has the same number of rows as the first matrix and the same number of columns as the second matrix. Also the number of columns in the first is the same as the number of rows in the second matrix. The value of an element in **C**(row i, column j) is determined by the general formula:

$$c_{i,j} = \sum_{k=1}^{n} a_{i,k} b_{k,j}$$

Thus,

$$\begin{bmatrix} 1 & 2 & 6 \\ 4 & 5 & 10 \\ 5 & 3 & 11 \end{bmatrix} \begin{bmatrix} 3 & 0 \\ 0 & 1 \\ 5 & 1 \end{bmatrix} = \begin{bmatrix} 1\times3+2\times0+6\times5 & 1\times0+2\times1+6\times1 \\ 4\times3+5\times0+10\times5 & 4\times0+5\times1+10\times1 \\ 5\times3+3\times0+11\times5 & 5\times0+3\times1+11\times1 \end{bmatrix} = \begin{bmatrix} 33 & 8 \\ 62 & 15 \\ 70 & 14 \end{bmatrix}$$

It can also be seen that multiplication of matrices is not commutative ($\mathbf{A}\,\mathbf{B} \neq \mathbf{B}\,\mathbf{A}$). Multiplication of a matrix by a scalar is done by multiplying each element by the scalar. $c\mathbf{A} = \mathbf{A}c = [caij]$

$$2\begin{bmatrix} 1 & 2 & 6 \\ 4 & 5 & 10 \\ 5 & 3 & 11 \end{bmatrix} = \begin{bmatrix} 2 & 4 & 12 \\ 8 & 10 & 20 \\ 10 & 6 & 22 \end{bmatrix}$$

Identity Matrix

The identity matrix is a special matrix whose elements are all zeroes except along the primary diagonal, which are occupied by ones. The identity matrix can be any size as long as the number of rows equals the number of columns.

$$I = \begin{bmatrix} 1 & 0 & 0 & 0 \\ 0 & 1 & 0 & 0 \\ 0 & 0 & 1 & 0 \\ 0 & 0 & 0 & 1 \end{bmatrix}$$

Determinant

The determinant is a property of any square matrix that describes the degree of coupling between equations. For a 2x2 matrix the determinant is:

$$\det(\mathbf{A}) = \begin{vmatrix} a & b \\ c & d \end{vmatrix} = ad - bc$$

Note that the vertical lines around the matrix elements denotes the determinant. For a 3x3 matrix the determinant is:

$$\det(\mathbf{A}) = \begin{vmatrix} a & b & c \\ d & e & f \\ g & h & i \end{vmatrix} = a\begin{vmatrix} e & f \\ h & i \end{vmatrix} - b\begin{vmatrix} d & f \\ g & i \end{vmatrix} + c\begin{vmatrix} d & e \\ g & h \end{vmatrix} = a(ei - fh) - b(di - fg) + c(dh - eg)$$

Larger matrices are computed in the same way where the element of the top row is multiplied by the determinant of matrix remaining once that element's row and column are removed. Terms where the top elements in odd columns are added and terms where the top elements in even rows are subtracted (assuming the top element is positive). For matrices larger than 3x3 however; it is probably quickest to use math software to do these calculations since they quickly become more complex with increasing size.

Solving for Eigenvalues and Eigenvectors

The eigenvalues (λ) and eigenvectors (**v**), are related to the **square** matrix **A** by the following equation. (Note: In order for the eigenvalues to be computed, the matrix must have the same number of rows as columns.

$$(\mathbf{A} - \lambda\mathbf{I}) \cdot \mathbf{v} = 0$$

This equation is just a rearrangement of the equation $\mathbf{A} \cdot \mathbf{v} = \lambda \cdot \mathbf{v}$ that was seen above. To solve this equation, the eigenvalues are calculated first by setting $\det(\mathbf{A}-\lambda\mathbf{I})$ to zero and then solving for λ. The determinant is set to zero in order to ensure non-trivial solutions for **v**, by a fundamental theorem of linear algebra.

$$A = \begin{bmatrix} 4 & 1 & 4 \\ 1 & 7 & 1 \\ 4 & 1 & 4 \end{bmatrix}$$

$$A - \lambda I = \begin{bmatrix} 4 & 1 & 4 \\ 1 & 7 & 1 \\ 4 & 1 & 4 \end{bmatrix} + \begin{bmatrix} -\lambda & 0 & 0 \\ 0 & -\lambda & 0 \\ 0 & 0 & -\lambda \end{bmatrix}$$

$$\det(A - \lambda I) = \begin{vmatrix} 4-\lambda & 1 & 4 \\ 1 & 7-\lambda & 1 \\ 4 & 1 & 4-\lambda \end{vmatrix} = 0$$

$$-54\lambda + 15\lambda^2 - \lambda^3 = 0$$

$$-\lambda(\lambda - 6)(\lambda - 9) = 0$$

$$\lambda = 0, 6, 9$$

For each of these eigenvalues, an eigenvector is calculated which will satisfy the equation (A-λI)v=0 for that eigenvalue. To do this, an eigenvalue is substituted into A-λI, and then the system of equations is used to calculate the eigenvector. For $\lambda = 6$

$$(\mathbf{A} - 6\mathbf{I})\,\mathbf{v} = \begin{bmatrix} 4-6 & 1 & 4 \\ 1 & 7-6 & 1 \\ 4 & 1 & 4-6 \end{bmatrix}\begin{bmatrix} x \\ y \\ z \end{bmatrix} = \begin{bmatrix} -2 & 1 & 4 \\ 1 & 1 & 1 \\ 4 & 1 & -2 \end{bmatrix}\begin{bmatrix} x \\ y \\ z \end{bmatrix} = 0$$

Using multiplication we get a system of equations that can be solved.

$$-2x + y + 4z = 0 \longrightarrow y = 2x - 4z \qquad (1)$$

$$x + y + z = 0 \longrightarrow y = -x - z \qquad (2)$$

$$4x + y - 2z = 0 \longrightarrow y = -4x + 2z \qquad (3)$$

Equating equations

$$2x - 4z = -4x + 2z$$

$$6x = 6z$$

$$x = z$$

Plugging this into the given equation

$$y = -x - x = -2x$$

There is one degree of freedom in the system of equations, so we have to choose a value for one variable. By convention we choose x = 1 then

$$x = 1$$

$$y = -2$$

$$z = 1$$

$$\mathbf{v} = \begin{bmatrix} 1 \\ -2 \\ 1 \end{bmatrix}$$

A degree of freedom always occurs because in these systems not all equations turn out to be independent, meaning two different equations can be simplified to the same equation. In this case a small number was chosen (x = 1) to keep the solution simple. However, it is okay to pick any number for x, meaning that each eigenvalue potentially has an infinite number of possible eigenvectors that are scaled based on the initial value of x chosen. Said another way, the eigenvector only points in a direction, but the magnitude of this pointer does not matter. For this example, getting an eigenvector that is

$$\mathbf{v} = \begin{bmatrix} 1 \\ -2 \\ 1 \end{bmatrix}$$

is identical to getting an eigenvector that is

$$\mathbf{v} = \begin{bmatrix} 2 \\ -4 \\ 2 \end{bmatrix}$$

or an eigenvector that is scaled by some constant, in this case 2.

Finishing the calcualtions, the same method is repeated for $\lambda = 0$ and $\lambda = 9$ to get their corresponding eigenvectors.

For $\lambda = 0$, $\mathbf{v} = \begin{bmatrix} -1 \\ 0 \\ 1 \end{bmatrix}$

For $\lambda = 9$, $\mathbf{v} = \begin{bmatrix} 1 \\ 1 \\ 1 \end{bmatrix}$

In order to check your answers you can plug your eigenvalues and eigenvectors back into the governing equation $\mathbf{A} \cdot \mathbf{v} = \lambda \cdot \mathbf{v}$. For this example, $\lambda = 6$ and $\mathbf{v} = \begin{bmatrix} 1 \\ -2 \\ 1 \end{bmatrix}$ was double checked.

$$\begin{bmatrix} 4 & 1 & 4 \\ 1 & 7 & 1 \\ 4 & 1 & 4 \end{bmatrix} \cdot \begin{bmatrix} 1 \\ -2 \\ 1 \end{bmatrix} = 6 \begin{bmatrix} 1 \\ -2 \\ 1 \end{bmatrix}$$

$$\begin{bmatrix} 4 - 2 + 4 \\ 1 - 14 + 1 \\ 4 - 2 + 4 \end{bmatrix} = \begin{bmatrix} 6 \\ -12 \\ 6 \end{bmatrix}$$

Therefore, $\lambda = 6$ and $\mathbf{v} = \begin{bmatrix} 1 \\ -2 \\ 1 \end{bmatrix}$ are both an eigenvalue-eigenvector pair

for the matrix $\mathbf{A} = \begin{bmatrix} 4 & 1 & 4 \\ 1 & 7 & 1 \\ 4 & 1 & 4 \end{bmatrix}$.

Calculating Eigenvalues and Eigenvectors using Numerical Software

Eigenvalues in Mathematica

For larger matrices (4x4 and larger), solving for the eigenvalues and eigenvectors becomes very lengthy. Therefore software programs like Mathematica are used. The example from the last section will be used to demonstrate how to use Mathematica. First we can generate the matrix **A**. This is done using the following syntax:

- A = {{4,1,4},{1,7,1},{4,1,4}}

It can be seen that the matrix is treated as a list of rows. Elements in the same row are contained in a single set of brackets and separated by commas. The set of rows are also contained in a set of brackets and are separated by commas. A screenshot of this is seen below. (Note: The "MatrixForm[]" command is used to display the matrix in its standard form. Also in Mathematica you must hit Shift + Enter to get an output.

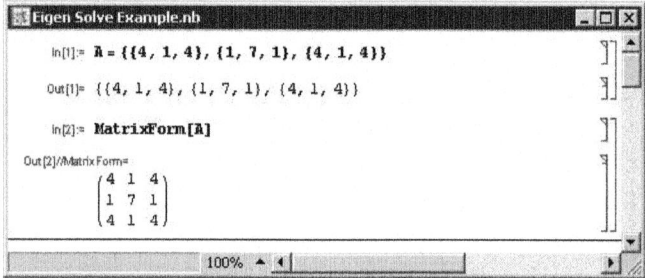

Next we find the determinant of matrix **A-λI**, by first subtracting the matrix λI from A (Note: This new matrix, **A-λI**, has been called A2).

In[9]:= i = {{1, 0, 0}, {0, 1, 0}, {0, 0, 1}}

Out[9]= {{1, 0, 0}, {0, 1, 0}, {0, 0, 1}}

In[10]:= MatrixForm[i]

Out[10]//MatrixForm=
$$\begin{pmatrix} 1 & 0 & 0 \\ 0 & 1 & 0 \\ 0 & 0 & 1 \end{pmatrix}$$

In[14]:= A2 = A - λ*i

Out[14]= {{4 - λ, 1, 4}, {1, 7 - λ, 1}, {4, 1, 4 - λ}}

In[15]:= MatrixForm[A2]

Out[15]//MatrixForm=
$$\begin{pmatrix} 4-\lambda & 1 & 4 \\ 1 & 7-\lambda & 1 \\ 4 & 1 & 4-\lambda \end{pmatrix}$$

The command to find the determinant of a matrix **A** is:

- Det[A]

For our example the result is seen below. By setting this equation to 0 and solving for λ, the eigenvalues are found. The Solve[] function is used to do this. Notice in the syntax that the use of two equal signs (==) is used to show equivalence whereas a single equal sign is used for defining a variable.

- Solve[{set of equations},{variables being solved}]

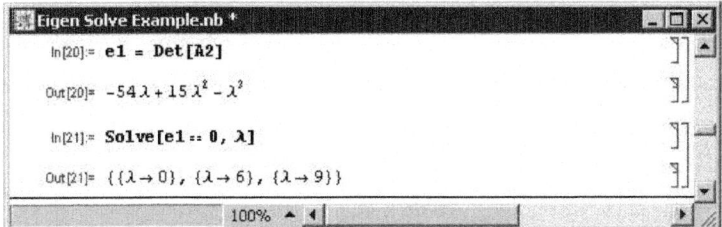

Alternatively the eigenvalues of a matrix **A** can be solved with the MathematicaEigenvalue[] function:

- Eigenvalues[A]

Note that the same results are obtained for both methods.

To find the eigenvectors of a matrix **A**, the Eigenvector[] function can be used with the syntax below.

- Eigenvectors[A]

The eigenvectors are given in order of descending eigenvalues.

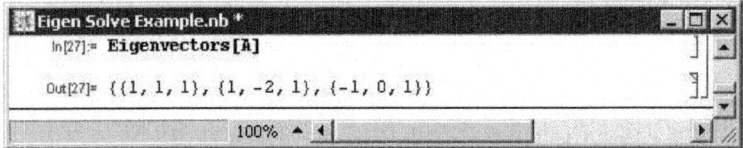

One more function that is useful for finding eigenvalues and eigenvectors is Eigensystem[]. This function is called with the following syntax.

- Eigensystem[A]

In this function, the first set of numbers are the eigenvalues, followed by the sets of eigenvectors in the same order as their corresponding eigenvalues.

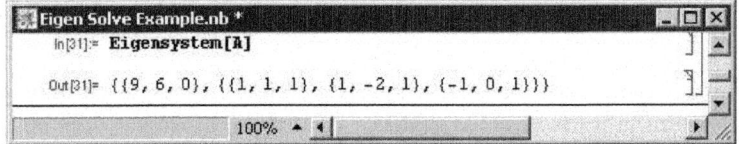

The Mathematica file used to solve the example.

Microsoft Excel

Microsoft Excel is capable of solving for Eigenvalues of symmetric matrices using its Goal Seek function. A symmetric matrix is a square matrix that is equal to its transpose and always has real, not complex, numbers for Eigenvalues. In many cases, complex Eigenvalues cannot be found using Excel. Goal Seek can be used because finding the Eigenvalue of a symmetric matrix is analogous to finding the root of a polynomial equation. The following procedure describes how to calculate the Eigenvalue of a symmetric matrix in the Mathematica tutorial using MS Excel.

(1) Input the values displayed below for matrix A then click menu INSERT-NAME-DEFINE "matrix_A" to name the matrix.

matrix_A		
4	1	4
1	7	1
4	1	4

This is the matrix.

(2) Similarly, define identity matrix I by entering the values displayed below then naming it "matrix_I."

matrix_I		
1	0	0
0	1	0
0	0	1

This is the identity matrix.

(3) Enter an initial guess for the Eigenvalue then name it "lambda."

lambda	6

(4) In an empty cell, type the formula =matrix_A-lambda*matrix_I. Highlight three cells to the right and down, press F2, then press CRTL+SHIFT+ENTER. Name this matrix "matrix_A_lambda_I."

matrix_A_lambda_I		
-2	1	4
1	1	1
4	1	-2

*This is matrix_A-lambda*I.*

(5) In another cell, enter the formula =MDETERM(matrix_A_lambda_I). This is the determinant formula for matrix_A_lambda_I.

det(matrix_A-lambda*I)	0

(6) Click menu Tools-Goal Seek… and set the cell containing the determinant formula to zero by changing the cell containing lambda.

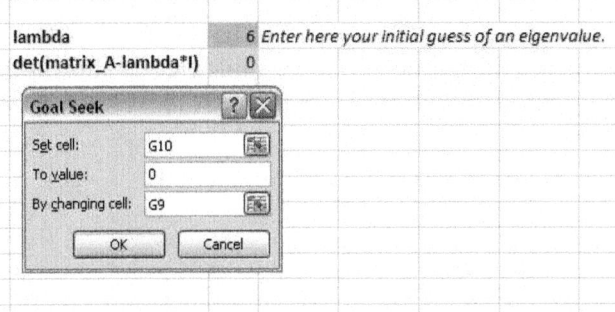

(7) To obtain all three Eigenvalues for matrix A, re-enter different initial guesses. Excel calculates the Eigenvalue nearest to the value of the initial guess. The Eigenvalues for matrix A were determined to be 0, 6, and 9. For instance, initial guesses of 1, 5, and 13 will lead to Eigenvalues of 0, 6, and 9, respectively.

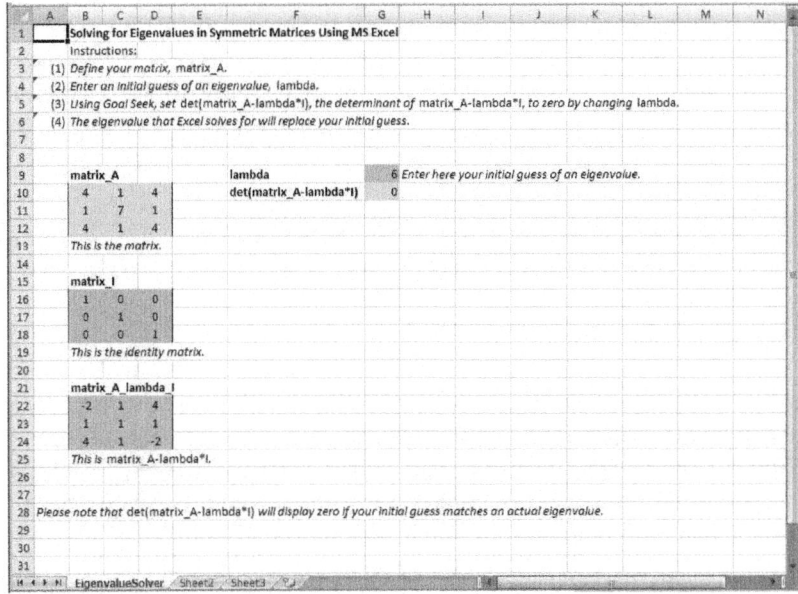

The MS Excel spreadsheet used to solve this problem.

Chemical Engineering Applications

The eigenvalue and eigenvector method of mathematical analysis is useful in many fields because it can be used to solve homogeneous linear systems of differential equations with constant coefficients. Furthermore, in chemical engineering many models are formed on the basis of systems of differential equations that are either linear or can be linearized and solved using the eigenvalue eigenvector method.

In general, most ODEs can be linearized and therefore solved by this method. Linearizing ODEs For example, a PID control device can be modeled with ODEs that may be linearized where the eigenvalue eigenvector method can then be implemented. If we have a system that can be modeled with linear differential equations involving temperature, pressure, and concentration as they change with time, then the system can be solved using eigenvalues and eigenvectors:

$$\frac{dP}{dt} = 4P - 4T + C$$

$$\frac{dT}{dt} = 4P - T + 3C$$

$$\frac{dC}{dt} = P + 5T - C$$

Note: This is not a real model and simply serves to introduce the eigenvalue and eigenvector method.

A is just the matrix that represents the coefficients in the above linear differential equations. However, when setting up the matrix, **A**, the order of coefficients matters and must remain consistent. Namely, in the following representative matrix, the first column corresponds to the coefficients of **P**, the second column to the coefficients of **T**, and the third column corresponds to the coefficients of **C**. The same goes for the rows. The first row corresponds to $\frac{dP}{dt}$, the second row corresponds to $\frac{dT}{dt}$, and the third row corresponds to $\frac{dC}{dt}$:

$$\mathbf{A} = \begin{bmatrix} 4 & -4 & 1 \\ 4 & -1 & 3 \\ 1 & 5 & -1 \end{bmatrix}$$

It is noteworthy that matrix **A** is only filled with constants for a linear system of differential equations. This turns out to be the case because each matrix component is the partial differential of a variable (in this case P, T, or C). It is this partial differential that yields a constant for linear systems. Therefore, matrix **A** is really the Jacobian matrix for a linear differential system.

Now, we can rewrite the system of ODE's above in matrix form.

$\mathbf{x}' = \mathbf{A}\mathbf{x}$

Where $\mathbf{x}(t) = \begin{bmatrix} P(t) \\ T(t) \\ C(t) \end{bmatrix}$

We guess trial solutions of the form

$\mathbf{x} = \mathbf{v}e^{\lambda t}$

since when we substitute this solution into the matrix equation, we obtain

$\lambda \mathbf{v}e^{\lambda t} = \mathbf{A}\mathbf{v}e^{\lambda t}$

After cancelling the nonzero scalar factor $e^{\lambda t}$, we obtain the desired eigenvalue problem.

$$\mathbf{A}\mathbf{v} = \lambda \mathbf{v}$$

Thus, we have shown that $\mathbf{x} = \mathbf{v}e^{\lambda t}$ will be a nontrivial solution for the matrix equation as long as $\mathbf{v}$ is a nonzero vector and λ is a constant associated with $\mathbf{v}$ that satisfies the eigenvalue problem.

In order to solve for the eigenvalues and eigenvectors, we rearrange the equation $\mathbf{A}\mathbf{v} = \lambda \mathbf{v}$ to obtain the following:

$$(\mathbf{A} - \lambda \mathbf{I})\mathbf{v} = 0 \quad \longrightarrow \quad \begin{bmatrix} 4 - \lambda & -4 & 1 \\ 4 & -1 - \lambda & 3 \\ 1 & 5 & -1 - \lambda \end{bmatrix} \cdot \begin{bmatrix} x \\ y \\ z \end{bmatrix} = 0$$

For nontrivial solutions for $\mathbf{v}$, the determinant of the eigenvalue matrix must equal zero, $det\,(\mathbf{A} - \lambda \mathbf{I}) = 0$. This allows us to solve for the eigenvalues, λ. You should get, after simplification, a third order polynomial, and therefore three eigenvalues. Using the calculated eignvalues, one can determine the stability of the system when disturbed.

Once you have calculated the three eigenvalues, you are ready to find the corresponding eigenvectors. Plug the eigenvalues back into the equation $(\mathbf{A} - \lambda \mathbf{I})\,\mathbf{v} = 0$ and solve for the corresponding eigenvectors. There should be three eigenvectors, since there were three eigenvalues.

The solution will look like the following:

$$\begin{bmatrix} P(t) \\ T(t) \\ C(t) \end{bmatrix} = c_1 \begin{bmatrix} x_1 \\ y_1 \\ z_1 \end{bmatrix} e^{\lambda_1 t} + c_2 \begin{bmatrix} x_2 \\ y_2 \\ z_2 \end{bmatrix} e^{\lambda_2 t} + c_3 \begin{bmatrix} x_3 \\ y_3 \\ z_3 \end{bmatrix} e^{\lambda_3 t}$$

Where $x_1, x_2, x_3, y_1, y_2, y_3, z_1, z_2, z_3$ are all constants from the three eigenvectors. The general solution is a linear combination of these three solution vectors because the original system of ODE's is homogeneous and linear. It is homogeneous because the derivative expressions have no cross terms, such as PC or TC, and no dependence on t. It is linear because the derivative operator is linear. To solve for c_1, c_2, c_3 there must be some given initial conditions.

This Wiki does not deal with solving ODEs. It only deals with solving for the eigenvalues and eigenvectors. In Mathematica the **Dsolve[]** function can be used to bypass the calculations of eigenvalues and eigenvectors to give the solutions for the differentials directly.

Using Eigenvalues to Determine Effects of Disturbing a System

Eigenvalues can help determine trends and solutions with a system of differential equations. Once the eigenvalues for a system are determined, the eigenvalues can be used to describe the system's ability to return to steady-state if disturbed.

The simplest way to predict the behaviour of a system if disturbed is to examine the signs of its eigenvalues. Negative eigenvalues will drive the system back to its steady-state value, while positive eigenvalues will drive it away. What happens if there are two eigenvalues present with opposite signs? How will the system respond to a disturbance in that case? In many situations, there will be one eigenvalue which has a much higher absolute value than the other corresponding eigenvalues for that system of differential equations.

This is known as the "dominant eigenvalue", and it will have the greatest effect on the system when it is disturbed. However, in the case that the eigenvalues are equal and opposite sign there is no dominant eigenvalue. In this case the constants from the initial conditions are used to determine the stability.

Another possible case within a system is when the eigenvalue is 0. When this occurs, the system will remain at the position to which it is disturbed, and will not be driven towards or away from its steady-state value. It is also possible for a system to have two identical eigenvalues. In this case the two identical eigenvalues produce only one eigenvector. Because of this, a situation can arise in which the eigenvalues don't give the complete story of the system, and another method must be used to analyze it, such as the Routh Stability Analysis Method.

Eigenvalues can also be complex or pure imaginary numbers. If the system is disturbed and the eigenvalues are non-real number, oscillation will occur around the steady state value. If the eigenvalue is imaginary with no real part present, then the system will oscillate with constant amplitude around the steady-state value. If it is complex with a positive real part, then the system will oscillate with increasing amplitude around the function, driving the system further and further away from its steady-state value. Lastly, if the eigenvalue is a complex number with a negative real part, then the system will oscillate with decreasing amplitude until it eventually reaches its steady state value again.

Below is a table of eigenvalues and their effects on a differential system when disturbed. It should be noted that the eigenvalues developed for a system should be reviewed as a system rather than as individual values. That is to say, the effects listed in the table below do not fully represent how the system will respond. If you were to pretend that eigenvalues were nails on a Plinko board, knowing the location and angle of one of those nails would not allow you to predict or know how the Plinko disk would fall down the wall, because you wouldn't know the location or angle of the other nails. If you have information about all of the nails on the Plinko board, you could develop a prediction based on that information.

Eigenvalue	Effect on system when disturbed
Positive real number	Driven away from steady-state value
Negative real number	Driven back to steady-state value
0	Remains at position to which it was disturbed
Identical to another eigenvalue	Effects can not be determined
Complex, positive real number	Oscillates around steady-state value with increasing amplitude
Complex, negative real number	Oscillates around steady-state value with decreasing amplitude
Imaginary	Oscillates around steady-state value with constant amplitude

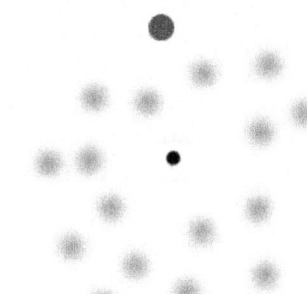

The above picture is of a plinko board with only one nail position known. Without knowing the position of the other nails, the Plinko disk's fall down the wall is unpredictable.

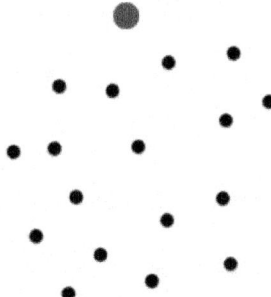

Knowing the placement of all of the nails on this Plinko board allows the player to know general patterns the disk might follow.

Repeated Eigenvalues

A final case of interest is repeated eigenvalues. While a system of N differential equations must also have N eigenvalues, these values may not always be distinct. For example, the system of equations:

$$\frac{dC_A}{dt} = f_{A\,in}\rho C_{A\,in} - f_{out}\rho C_A \sqrt{V_1} - V_1 k_1 C_A C_B$$

$$\frac{dC_B}{dt} = f_{B\,in}\rho C_{B\,in} - f_{out}\rho C_B \sqrt{V_1} - V_1 k_1 C_A C_B$$

$$\frac{dC_C}{dt} = -f_{out}\rho C_C \sqrt{V_1} + V_1 k_1 C_A C_B$$

$$\frac{dV_1}{dt} = f_{A\,in} + f_{B\,in} - f_{out}\sqrt{V_1}$$

$$\frac{dV_2}{dt} = f_{out}\sqrt{V_1} - f_{customer}\sqrt{V_2}$$

$$\frac{dC_{C2}}{dt} = f_{out}\rho C_C \sqrt{V_1} - f_{customer}\rho C_{C2} \sqrt{V_2}$$

May yield the eigenvalues: {-82, -75, -75, -75, -0.66, -0.66}, in which the roots '-75' and '-0.66' appear multiple times. Repeat eigenvalues bear further scrutiny in any analysis because they might represent an edge case, where the system is operating at some extreme. In mathematical terms, this means that linearly independent eigenvectors cannot be generated to complete the matrix basis without further analysis. In "real-world" engineering terms, this means that a system at an edge case could distort or fail unexpectedly.

However, for the general solution:

$$Y(t) = k_1 \exp(\lambda t)V_1 + k_2 \exp(\lambda t)(t V_1 + V_2)$$

If $\lambda < 0$, as t approaches infinity, the solution approaches 0, indicating a stable sink, whereas if $\lambda > 0$, the solution approaches infinity in the limit, indicating an unstable source. Thus the rules above can be roughly applied to repeat eigenvalues, that the system is still likely stable if they are real and less than zero and likely unstable if they are real and positive. Nonetheless, one should be aware that unusual behaviour is possible.

STABILITY

Eigenvalues can be used to determine whether a fixed point (also known as an equilibrium point) is stable or unstable. A stable fixed point is such that a system can be initially disturbed around its fixed point yet eventually return to its original location and remain there. A fixed point is unstable if it is not stable. To illustrate this concept, imagine a round ball in between two hills. If left alone, the ball will not move, and thus its position is considered a fixed point. If we were to disturb the ball by pushing it a little bit up the hill, the ball will roll back to its original position in between the two hills. This is a stable fixed point. Now image that the ball is at the peak of one of the hills. If left undisturbed, the ball will still remain at the peak, so this is also considered a fixed point. However, a disturbance in any direction will cause the ball to roll away from the top of the hill. The top of the hill is considered an unstable fixed point.

The eigenvalues of a system linearized around a fixed point can determine the stability behaviour of a system around the fixed point. The particular stability behaviour depends upon the existence of real and imaginary components of the eigenvalues, along with the signs of the real components and the distinctness of their values. We will examine each of the possible cases below.

Imaginary (or Complex) Eigenvalues

When eigenvalues are of the form $a + bi$, where a and b are real scalars and i is the imaginary number $\sqrt{-1}$, there are three important cases. These three cases are when the real part a is positive, negative, and zero. In all cases, when the complex part of an eigenvalue is non-zero, the system will be oscillatory.

Positive Real Part

When the real part is positive, the system is unstable and behaves as an unstable oscillator. This can be visualized as a vector tracing a spiral away from the fixed point. The plot of response with time of this situation would look sinusoidal with ever-increasing amplitude, as shown below.

This situation is usually undesirable when attempting to control a process or unit. If there is a change in the process, arising from the process itself or from an external disturbance, the system itself will not go back to steady state.

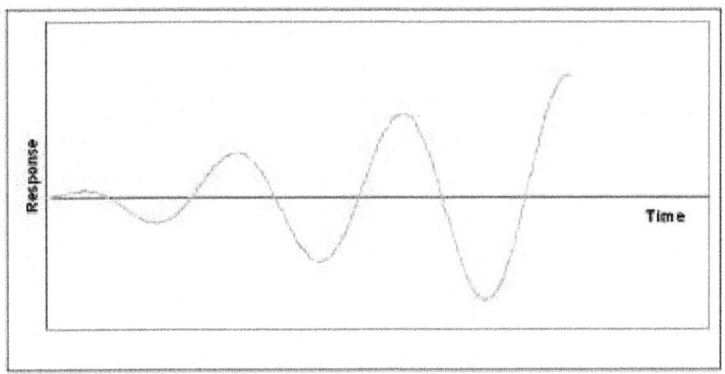

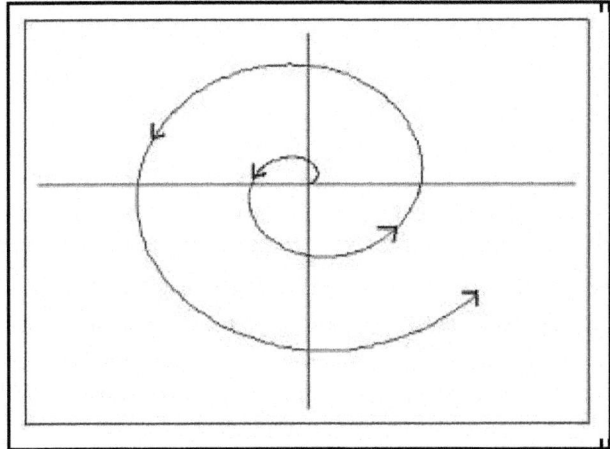

Zero Real Part

When the real part is zero, the system behaves as an undamped oscillator. This can be visualized in two dimensions as a vector tracing a circle around a point. The plot of response with time would look sinusoidal. The figures below should help in understanding.

Undamped oscillation is common in many control schemes arising out of competing controllers and other factors. Even so, this is usually undesirable and is considered an unstable process since the system will not go back to steady state following a disturbance.

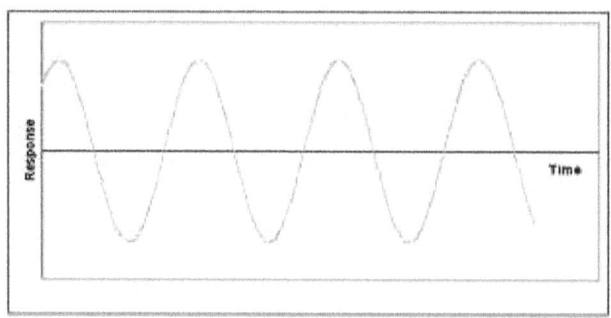

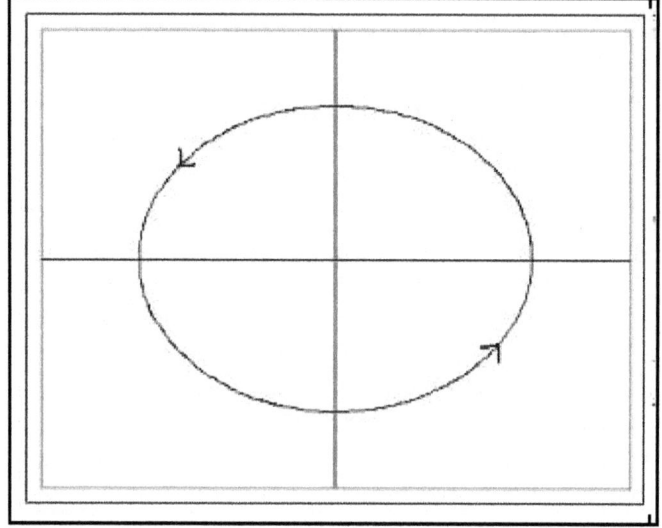

Negative Real Part

When the real part is negative, then the system is stable and behaves as a damped oscillator. This can be visualized as a vector tracing a spiral toward the fixed point. The plot of response with time of this situation would look sinusoidal with ever-decreasing amplitude, as shown below.

This situation is what is generally desired when attempting to control a process or unit. This system is stable since steady state will be reached even after a disturbance to the system. The oscillation will quickly bring the system back to the setpoint, but will over shoot, so if overshooting is a large concern, increased damping would be needed.

While discussing complex eigenvalues with negative real parts, it is important to point out that having all negative real parts of eigenvalues is a necessary and sufficient condition of a stable system.

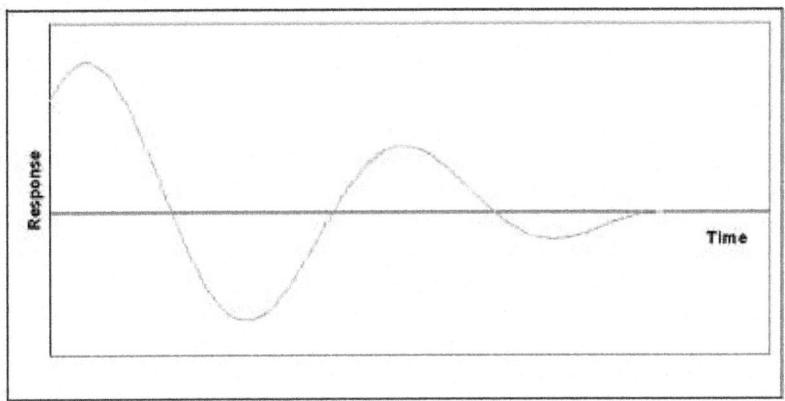

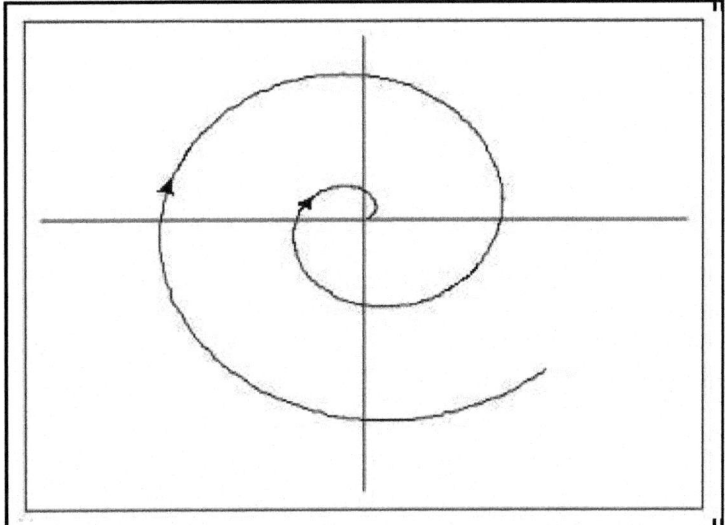

Complex Part of Eigenvalues

As previously noted, the stability of oscillating systems (*i.e.* systems with complex eigenvalues) can be determined entirely by examination of the real part. Although the sign of the complex part of the eigenvalue may cause a phase shift of the oscillation, the stability is unaffected.

Real Eigenvalues

We've seen how to analyze eigenvalues that are complex in form, now we will look at eigenvalues with only real parts.

Zero Eigenvalues

If an eigenvalue has no imaginary part and is equal to zero, the system will be unstable, since, as mentioned earlier, a system will not be stable if its eigenvalues have any non-negative real parts. This is just a trivial case of the complex eigenvalue that has a zero part.

Positive Eigenvalues

When all eigenvalues are real, positive, and distinct, the system is unstable. On a gradient field, a spot on the field with multiple vectors circularly surrounding and pointing out of the same spot (a node) signifies all positive eigenvalues. This is called a source node.

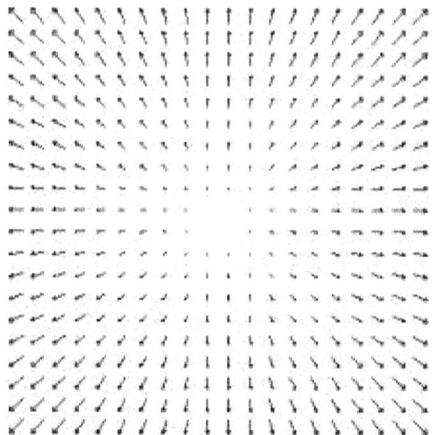

Graphically, real and positive eigenvalues will show a typical exponential plot when graphed against time.

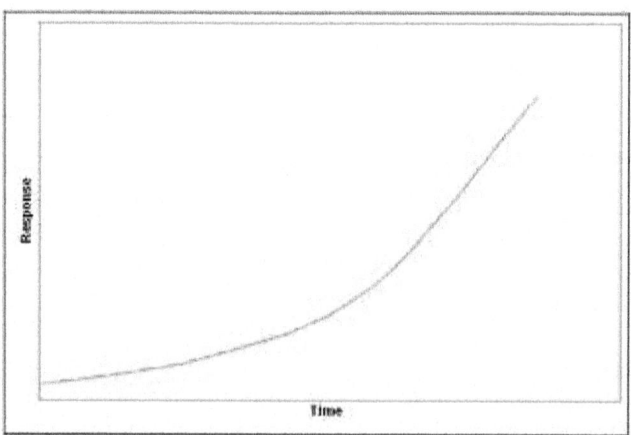

Negative Eigenvalues

When all eigenvalues are real, negative, and distinct, the system is stable. Graphically on a gradient field, there will be a node with vectors pointing toward the fixed point. This is called a sink node.

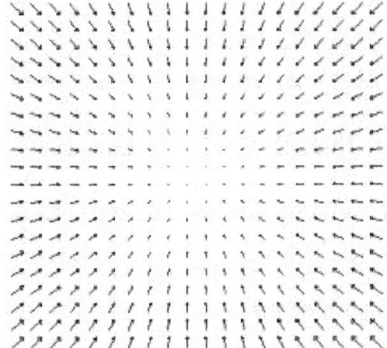

Graphically, real and negative eigenvalues will output an inverse exponential plot.

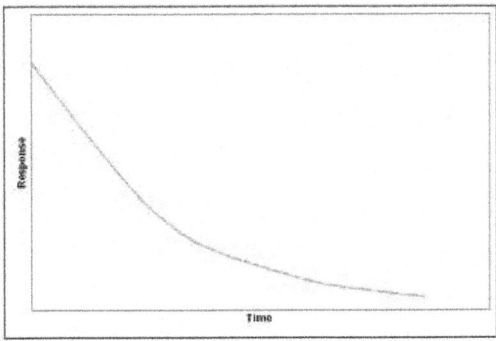

Positive and Negative Eigenvalues

If the set of eigenvalues for the system has both positive and negative eigenvalues, the fixed point is an unstable saddle point. A saddle point is a point where a series of minimum and maximum points converge at one area in a gradient field, without hitting the point. It is called a saddle point because in 3 dimensional surface plot the function looks like a saddle.

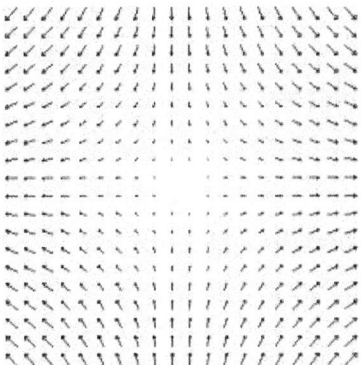

Repeated Eigenvalues

If the set of eigenvalues for the system has repeated real eigenvalues, then the stability of the critical point depends on whether the eigenvectors associated with the eigenvalues are linearly independent, or orthogonal. This is the case of degeneracy, where more than one eigenvector is associated with an eigenvalue. In general, the determination of the system's behaviour requires further analysis. For the case of a fixed point having only *two* eigenvalues, however, we can provide the following two possible cases. If the two repeated eigenvalues are positive, then the fixed point is an unstable source. If the two repeated eigenvalues are negative, then the fixed point is a stable sink.

Summary of Eigenvalue Graphs

Below is a table summarizing the visual representations of stability that the eigenvalues represent.

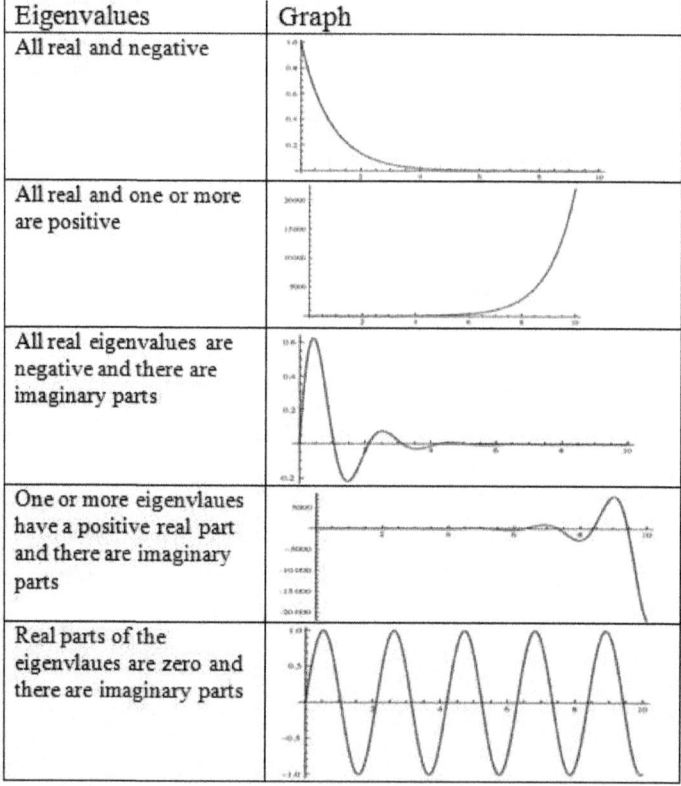

Eigenvalues	Graph
All real and negative	
All real and one or more are positive	
All real eigenvalues are negative and there are imaginary parts	
One or more eigenvlaues have a positive real part and there are imaginary parts	
Real parts of the eigenvlaues are zero and there are imaginary parts	

Another Method of Determining Stability

The process of finding eigenvalues for a system of linear equations can become rather tedious at times and to remedy this, a British mathematician named Edward Routh came up with a handy little short-cut.

First, recall that an unstable eigenvalue will have a positive or zero real part and that a stable eigenvalue will have a negative real part.

The first test is to take an n-th degree polynomial of interest:

$$P(\lambda) = a_0\lambda^n + a_1\lambda^{n-1} + \cdots + a_{n-1}\lambda + a_n$$

and look to see if any of the coefficients are negative or zero. If so, there is at least one value with a positive or zero real part which refers to an unstable node.

The way to test exactly how many roots will have positive or zero real parts is by performing the complete Routh array. Referring to the previous polynomial, it works as follows:

Row					
1	a_0	a_2	a_4	a_6	$\cdots$
2	a_1	a_3	a_5	a_7	$\cdots$
3	b_1	b_2	b_3	b_4	$\cdots$
4	c_1	c_2	c_3	c_4	$\cdots$
$\vdots$	$\vdots$	$\vdots$	$\vdots$	$\vdots$	
n-1	p_1	p_2			
n	q_1				
n+1	v_1				

An array of n+1 rows and the coefficients placed as above. After the first two rows, the values are obtained as below:

$$b_1 = \frac{a_1 a_2 - a_0 a_3}{a_1}, b_2 = \frac{a_1 a_4 - a_0 a_5}{a_1}, b_3 = \frac{a_1 a_6 - a_0 a_7}{a_1}, \cdots$$

$$c_1 = \frac{b_1 a_3 - a_1 b_2}{b_1}, c_2 = \frac{b_1 a_5 - a_1 b_3}{b_1}, c_3 = \frac{b_1 a_7 - a_1 b_4}{b_1}, \cdots$$

Routh's theorem says:

1. For all of the roots of the polynomial to be stable, all the values in the first column of the Routh array must be positive.

2. If any of the values in the first column are negative, then the number of roots with a positive real part equals the number of sign changes in the first column.

So considering the following example,

$$f(x) = 9x^4 + 14x^3 + 7x + 10$$

Preliminary test: All of the coefficients are positive, however, there is a zero coefficient for x^2 so there should be at least one point with a negative or zero real part.

Routh array:

Row			
1	9	0	10
2	14	7	
3	−4.5	10	
4	38.1		
5	10		

Since Row 3 has a negative value, there is a sign change from Row 2 to Row 3 and again from Row 3 to Row 4. Thus, there are 2 roots with positive or zero real part.

Stability Summary

The following image can work as a quick reference to remind yourself of what vector field will result depending on the eigenvalue calculated.

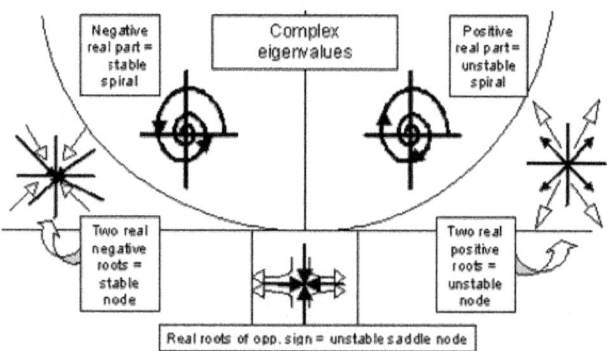

The table below gives a complete overview of the stability corresponding to each type of eigenvalue.

Eigenvalue Type	Stability	Oscillatory Behaviour	Notation
All Real and +	Unstable	None	Unstable Node
All Real and -	Stable	None	Stable Node
Mixed + & - Real	Unstable	None	Unstable saddle point
+a + bi	Unstable	Undamped	Unstable spiral
-a + bi	Stable	Damped	Stablespriral
0 + bi	Unstable	Undamped	Circle
Repeated values	Depends on orthogonality of eigenvectors		

Advantages and Disadvantages of Eigenvalue Stability

There are several advantages of using eigenvalues to establish the stability of a process compared to trying to simulate the system and observe the results. However, there are situations where eigenvalue stability can break down for some models.

Advantages

1. High accuracy for linear systems.
2. General method that can be applied to a variety of processes.
3. Can be used even if all variables are not defined, such as control parameters.

Disadvantages

1. Only applicable for linear models.
2. Linear approximations of nonlinear models break down away from the fixed point of approximation.

Chapter 3

PHASE PLANE ANALYSIS

INTRODUCTION TO ATTRACTORS, SPIRALS AND LIMIT CYCLES

We often use differential equations to model a dynamic system such as a valve opening or tank filling. Without a driving force, dynamic systems would stop moving. At the same time dissipative forces such as internal friction and thermodynamic losses are taking away from the driving force. Together the opposing forces cancel any interruptions or initial conditions and cause the system to settle into typical behaviour. Attractors are the location that the dynamic system is drawn to in its typical behaviour. Attractors can be fixed points, limit cycles, spirals or other geometrical sets.

Limit cycles are much like sources or sinks, except they are closed trajectories rather than points. Once a trajectory is caught in a limit cycle, it will continue to follow that cycle. By definition, at least one trajectory spirals into the limit cycle as time approaches either positive or negative infinity. Like a sink, attractive (stable) limit cycles have the neighboring trajectories approaching the limit cycle as time approaches positive infinity. Like a source, non-attractive (unstable) limit cycles have the neighboring trajectories approaching the limit cycle as time approaches negative infinity. Below is an illustration of a limit cycle.

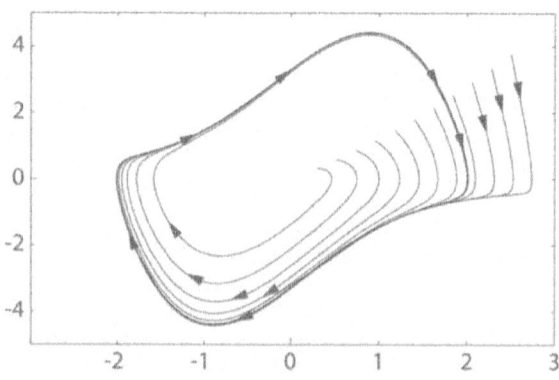

Spirals are a similar concept. The attractor is a spiral if it has complex eigenvalues. If the real portion of the complex eigenvalue is positive (*i.e.* 3 + 2i), the attractor is unstable and the system will move away from steady-state operation given a disturbance. If the real portion of the eigenvalue is negative (*i.e.* -2 + 5i), the attractor is stable and will return to steady-state operation given a disturbance.

Given the following set of linear equations we will walk through an example that produces a spiral:

$$\frac{dx}{dt} = 2x + 5y$$

$$\frac{dy}{dt} = -5x + 2y$$

The Jacobian matrix would be the coefficients:

$$A = \begin{vmatrix} 2 & 5 \\ -5 & 2 \end{vmatrix}$$

Next we found the eigenvalues:

$$(A - \lambda I) = \begin{vmatrix} 2 & 5 \\ -5 & 2 \end{vmatrix} - \lambda \begin{vmatrix} 1 & 0 \\ 0 & 1 \end{vmatrix} = \begin{vmatrix} (2 - \lambda) & 5 \\ -5 & (2 - \lambda) \end{vmatrix}$$

where I is the identity matrix $\begin{vmatrix} 1 & 0 \\ 0 & 1 \end{vmatrix}$

$$det(A - \lambda I) = (2 - \lambda)^2 + 25 = 0$$

Eigenvalues: $\lambda = 2 \pm 5i$

The system is unstable because the real portion of the complex eigenvalues is positive.

To find the first eigenvector we continue by plugging in $2 - 5i$:

$$\begin{vmatrix} (2 - \lambda) & 5 \\ -5 & (2 - \lambda) \end{vmatrix} = \begin{vmatrix} 2 - (2 - 5i) & 5 \\ -5 & 2 - (2 - 5i) \end{vmatrix} = \begin{vmatrix} 5i & 5 \\ -5 & 5i \end{vmatrix}$$

$$(A - \lambda I)v = \begin{vmatrix} 5i & 5 \\ -5 & 5i \end{vmatrix} v = 0$$

let $v = \begin{vmatrix} x \\ y \end{vmatrix}$

$$\begin{vmatrix} 5i & 5 \\ -5 & 5i \end{vmatrix} \begin{vmatrix} x \\ y \end{vmatrix} = \begin{vmatrix} 0 \\ 0 \end{vmatrix}$$

We now have a system of equations which we can solve for x, y:

$5\,ix + 5y = 0$

$-5\,x + 5iy = 0$

Dividing both equations by 5:

$ix + y = 0$

$-x + iy = 0$

Solution: $v_1 = \begin{vmatrix} -1 \\ i \end{vmatrix}$

Following the same procedure using the second eigenvalue of $2 + 5i$, we find the second eigenvector to be:

$$v_2 = \begin{vmatrix} i \\ -1 \end{vmatrix}$$

Now plugging both eigenvalues and eigenvectors into the characteristic equation:

$$x(t) = e^{2t}(C_1 cos5t + C_2 sin5t)$$

$$y(t) = e^{2t}(C_3 cos5t + C_4 sin5t)$$

The phase-plane plot is shown below:

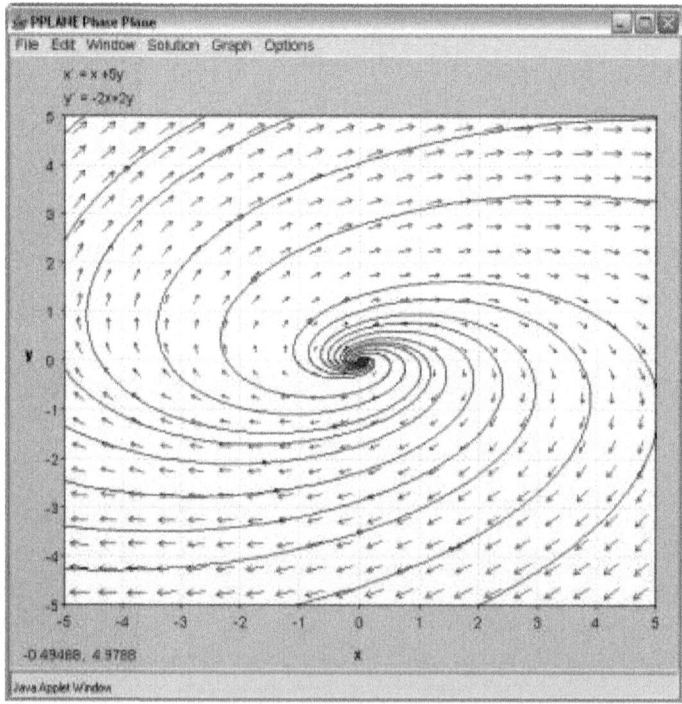

INTRODUCTION TO PPLANE

Phase-plane analysis is an important tool in studying the behaviour of nonlinear systems since there is often no analytical solution for a nonlinear system model.

PPlane is a JAVA applet for phase plane analysis of two-dimensional systems. It starts in your web browser and you can directly input your equations and parameter values. PPLANE plots vector fields for systems of differential equations. At each point, (x,y), of a grid, PPLANE draws an arrow indicating the direction and magnitude of the vector (x′,y′). This vector equals

$$dy/dt/ \; dx/dt = dy/dx,$$

and is independent of t; therefore, it must be tangent to any solution curve through (x,y).

It allows the user to plot solution curves in the phase plane by simple clicking on them. It also enables the user to plot these solutions in a variety of plots. There are a number of advanced features, including finding equilibrium points, eigenvalues and nullclines, that you will find useful later.

How to use Pplane

In the PPlane equation window you can enter a system of differential equations of the form

$$dx/dt = f(x,y) \text{ and } dy/dt = g(x,y),$$

define parameters and resize the display window. Under the Gallery pull down from the menu, you can switch to a linear system.

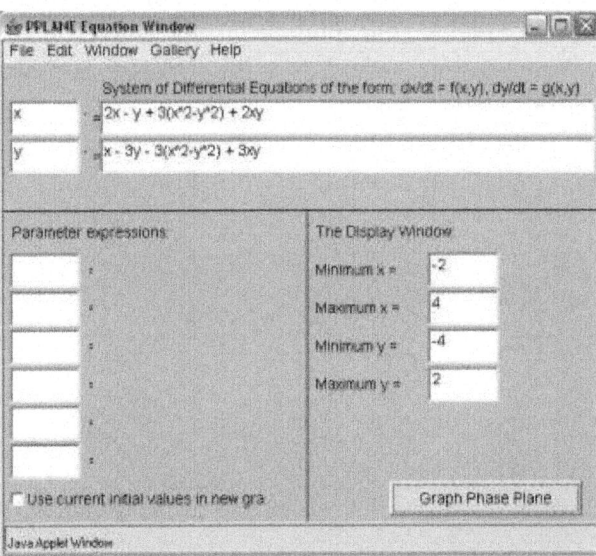

Note, if your differential equations contain constant parameters, you can enter them in the "Parameter Expressions" boxes below the differential equations as seen in the figure below (A, B, and C are used as example parameters). This is a convenient feature to use when considering the effect of changed parameters on the steady state of a system because it eliminates the redundancy of re-entering the parameter values multiple times within the differential equations.

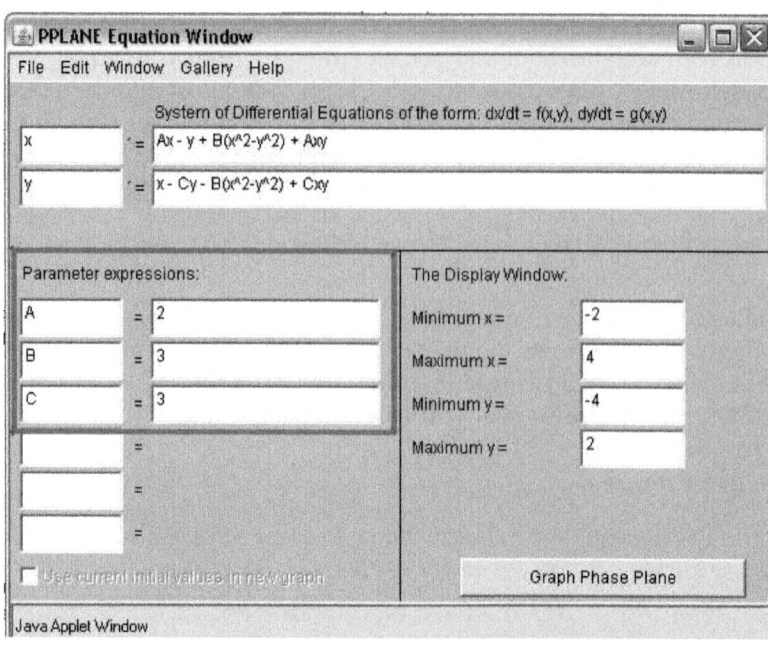

In the PPlane Phase Plane window below you will see the vector fields for the system. By clicking on the field you will plot solution curves in the phase plane. If you are interested in a plot of your solution vs. time or a 3-D view, click on graph:

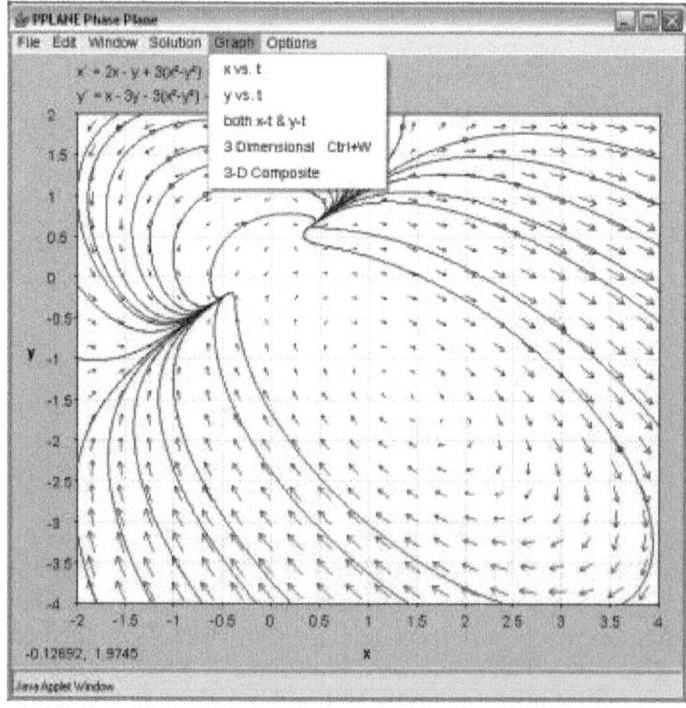

If you choose the x-t and y-t option, you have to pick a specific solution curve. The result will look like this:

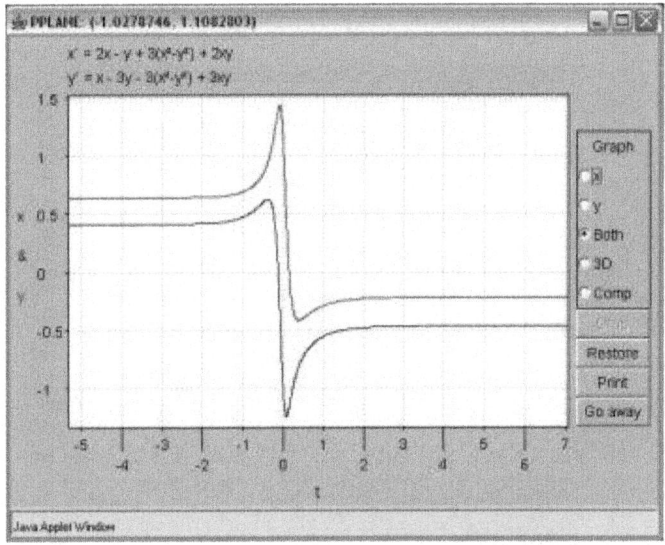

Use the crop function to zoom in on a point of interest

Choose Find Equillibrium Point under the Solution pull down menu. Then when you click on an orbit in the phaseplane, the Pplane Messages window will display the eigenvalues and possible equillibrium points.

Additional Things you can Change in PPlane

Changing the Slope Field

By clicking on the "Options" tab, then by selecting "Direction Field Settings," you can change the number of rows and columns plotted, the way the field is made up, as well as the computational settings of PPlane.

Erasing Made Orbits

On the "Edit" tab, there are options that say "Delete Orbit" or "Delete All Orbits". These options act as their names imply.

Changing the Direction of Graphing

By clicking on the "Options" tab, then by selecting the "Solution Direction" option, you can then change the way that PPlane graphs a line when you click on the field. You can change the graphing to plot forwards (for values of t>0), backwards (for values of t<0) or in both directions of t.

More Uses for PPLANE

Of the many uses that PPLANE has to offer, some of the most helpful functions involve:

Finding Eigenvalues/Eigenvectors for an Equilibrium Point

After graphing a series of differential equations, find a particular equilibrium point within the graphed data ("Find an Equilibrium Point," under the Solution tab)

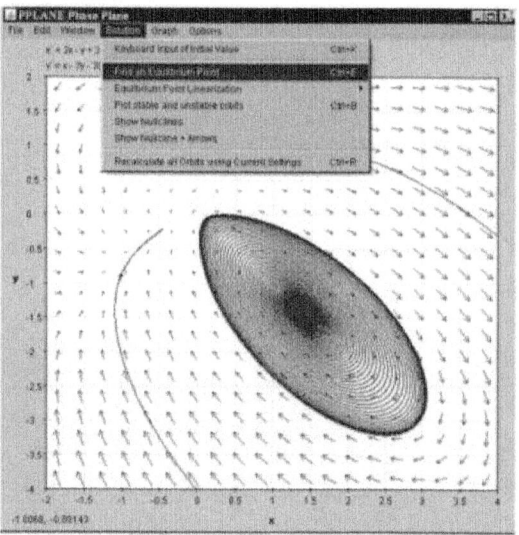

Now, by selecting a point on the field that has ben graphed by pplane, pplane will find the closest equilibrium point on the graph, and highlight this point on the graph in red.

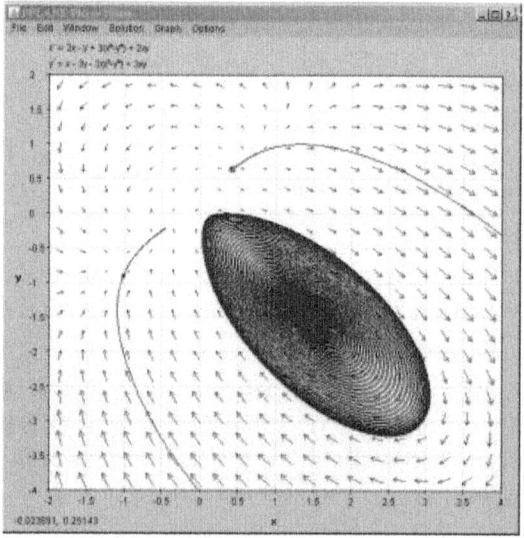

The PPLANE Messages box in the upper left hand corner of the screen should pop up with some new information. This information provides eigenvalues and the corresponding eigenvectors to the selected equilibrium value:

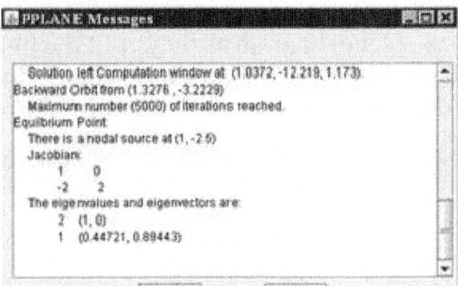

Stability of a Equilibrium Point

Similarly to before, the Messages Box will provide the stability features (ie: is it a nodal sink?) of the found equilibrium point, immediately after using the "Find an Equilibrium Point."

Other Concepts of Phase Plane Analysis

Separatrix

A separatrix is any line in the phase-plane that is not crossed by any trajectory. The unstable equilibrium point, or saddle point, below illustrates the idea of a separatrix, as neither the x or y axis is crossed by a trajectory. If you picture a topographic map, the seperatrix would be a mountain ridge; if you fall a little of the edge, you will never come back. Plotting your phase plane in Pplane would be useful to identify impossible set points, for example.

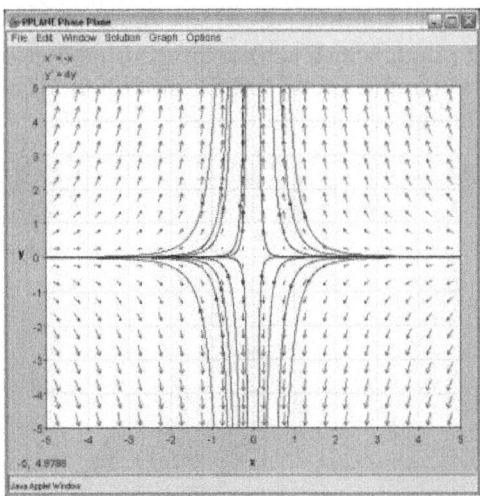

Nullclines

A nullcline is a curve where x′=0 or y′=0, thus indicating where the phase plane is completely horizontal or completely vertical. The point at which two nullclines intersect is an equilibrium point. Nullclines can also be quite useful for visualization of a phase plane diagram as they split the phase plane into regions of similar flow. To display nullclines on the Phase Plane window, select Nullclines under the Solutions drop down menu. The screenshot below is an example.

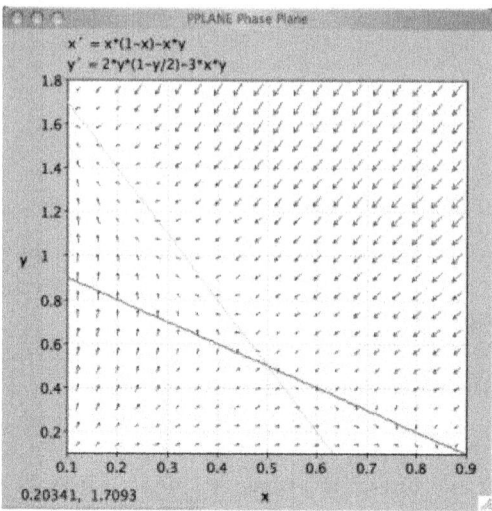

Notice that the red nullcline shows where the flow is completely vertical ($x'=0$) and the yellow nullcline shows where the flow is completely horizontal ($y'=0$).

Limit Cycle

Below you will find a solution curve for a limit cycle. The limit cycle contains the response in a set range, which is something you may want to take advantage for certain engineering applications. On the other hand it is always rotating and may not be stable enough for your purposes.

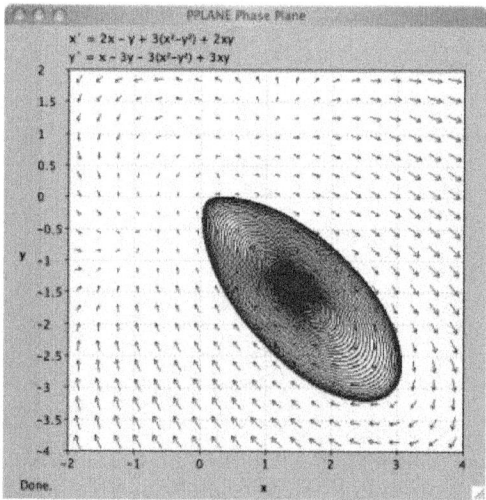

Taking Screen Shots to Copy Pplane Phase Portraits

With the introduction of Windows Vista, the Snipping Tool was introduced. This tool allows much greater flexibility with taking screen shots and editing them. This article will talk about the Snipping Tool as well as the Windows Print Screen key which can be used to take photos of your computer screen. When pressing the key, your computer copies the image of your screen and onto your computer's clipboard. The image can then be pasted into multiple programs.

To enable the Snipping Tool on your Vista computer go to the Windows button in the bottom left of your screen and click Accessories → Snipping Tool.

Fig. : How to enable the Snipping Tool.

A window will appear asking if you would like to add the Snipping Tool to your Quicklaunch. This provides a simple and quick way to take screenshots.

To take a picture of your graph, just press the Snipping Tool button in the Quicklaunch area and a window like this will appear:

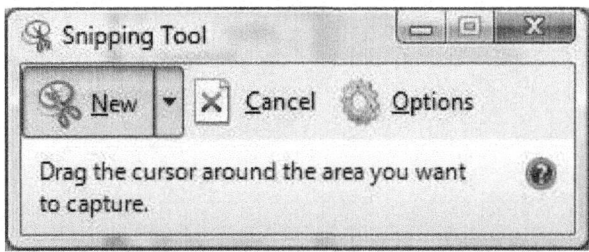

Fig. : The Snipping Tool Window.

Automatically, the Snipping Tool will default to a crosshair from which you can click and drag to make a selection of the section of the screen you would like represented by a red rectangle.

WARNING: In the Options section you should uncheck "Show selection ink after snips are captured" in order to eliminate the red edge around your photos.

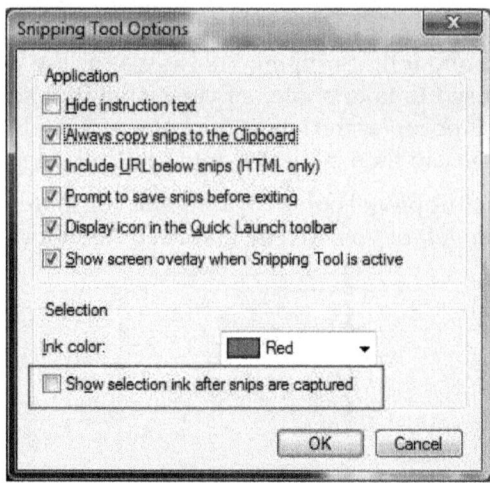

Fig. : Snipping Tool Option Menu (Uncheck the selection ink).

The Snipping Tool will open up a new window with your selection and copy the image to your clipboard. Feel free to edit your image or save it where it is convenient.

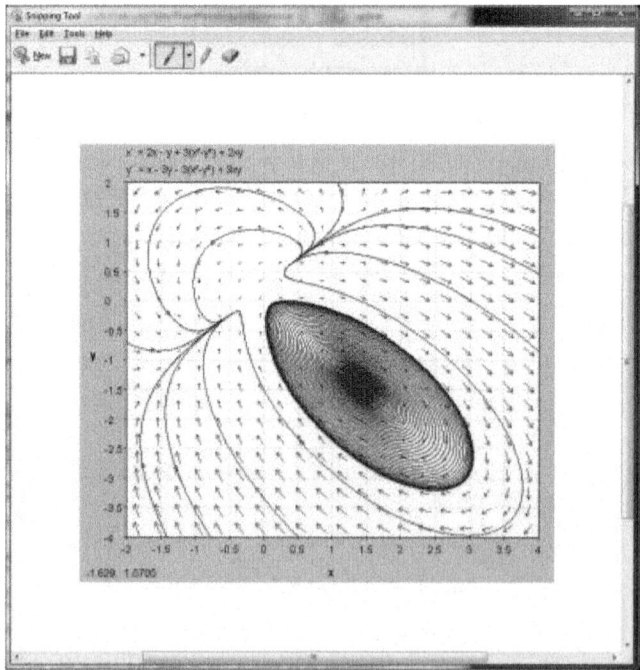

Fig. : Snipping Tool Editing Window.

If not using Windows Vista you can still use Print Screen:

Follow these simple steps to copy and paste your phase portrait into a Microsoft Word document:

1. Pull up the window containing your phase portrait so that it is displayed on the screen.

2. Find the Print Screen or PrtSc button in the upper-right hand portion of your keyboard. (The key may appear slightly different depending on your Windows keyboard manufacturer).

3. Open Microsoft Word to the document of your choice.

4. Paste the image into the Word document. Figure below indicates how your phase portrait will look in Word.

5. To crop or resize the image as you like, you may use the Picture toolbar by selecting View → Toolbars →Picture.

If you prefer to take a screen shot of just your phase portrait rather than the entire computer screen, follow these simple steps:

1. Pull up the window containing your phase portrait so that it is displayed on the screen.

2. Press Alt-Print Screen to capture a photo of the window you selected.

3. Open Microsoft Word to the document of your choice.

4. Paste the image into the Word document. Figure below indicates how your phase portrait image will look.

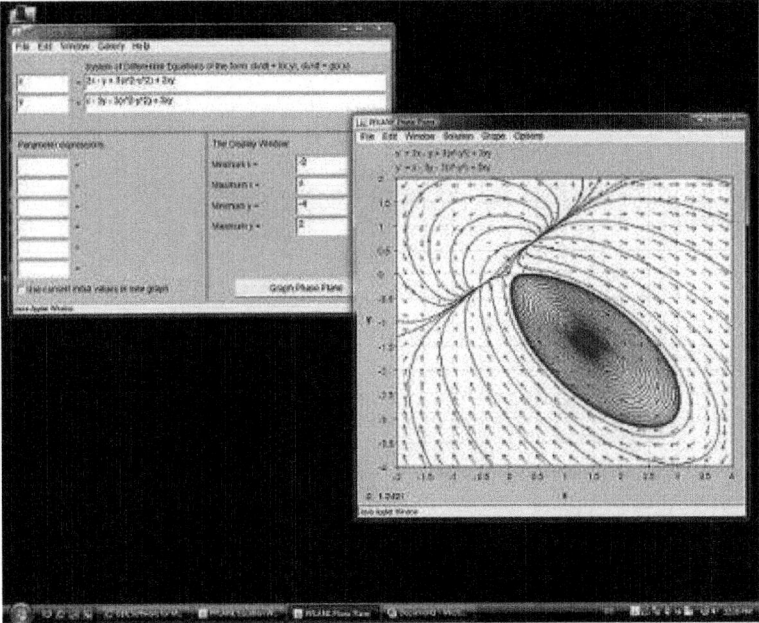

Fig. : Initial screen shot.

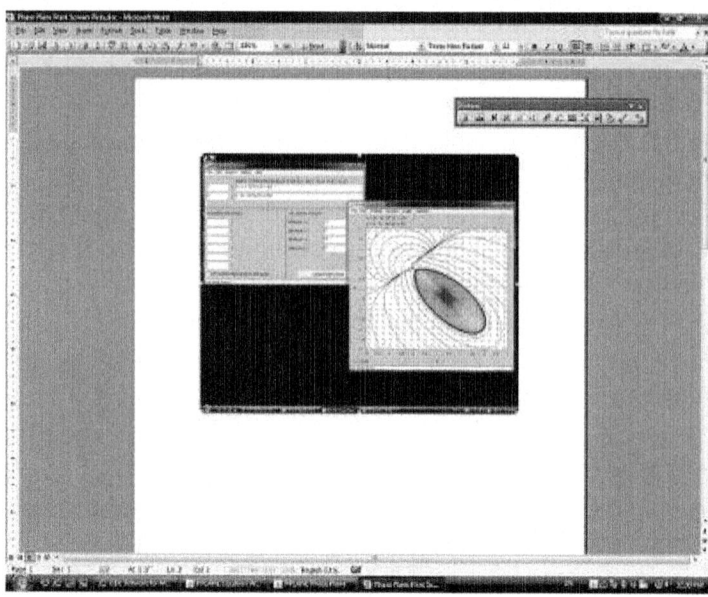

Fig. : Microsoft Word document containing the screen shot.

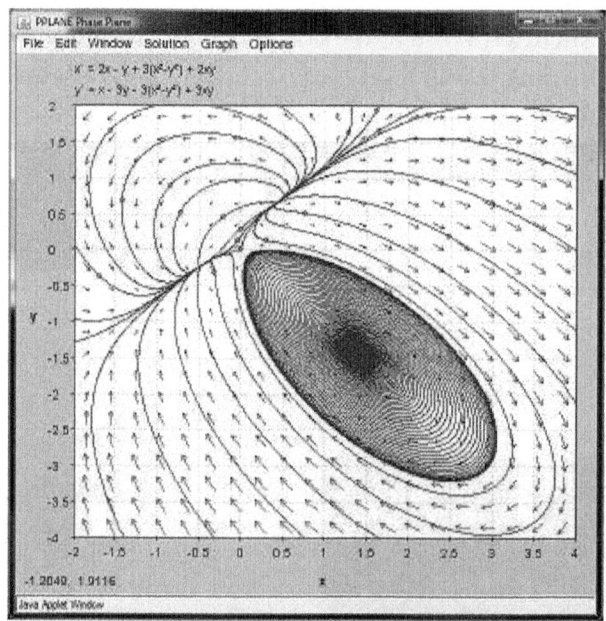

Fig. : Screen shot of phase portrait window.

ROOT LOCUS PLOTS

Root locus plots show the roots of the systems characteristic equation, (*i.e.* the Laplacian), as a function of the control variables such as Kc. By examining these graphs it is possible to determine the stability of different values of the control variable. A typical transfer function is of the form $G(s) = Y(s)/U(s)$.

Poles: $U(s) = 0$

Zeros: $Y(s) = 0$

In other words, after factorization the poles are the roots of the denominators and the zeros are the roots of the numerator. Stability only depends on the value of the poles. The system is stable for all values of the control variables that result in the value of the real part of every pole value being less than zero. The lines of a Root locus plot display the poles for values of the control variable(s) from zero to infinity on complex coordinate system. These plots will always have a line of symmetry at $i = 0$.

Closed-loop *vs.* Open-loop

A closed-loop system uses feedback control where the output has an effect on the input. With a closed-loop, oscillations are usually introduced, and therefore can become unstable. Unlike the open-loop systems, the closed-loop incorporates valves and controllers.

In an open-loop system, the output is not compared with and has no effect on the input. In an open-loop system, oscillations are not introduced and therefore cannot become unstable. An open-loop system, however, can be inaccurate because it does not take into account the control dynamics. Open-loop systems will include feed forward control schemes or timed control schemes. The two diagrams below depict this difference when trying to control the temperature in your apartment.

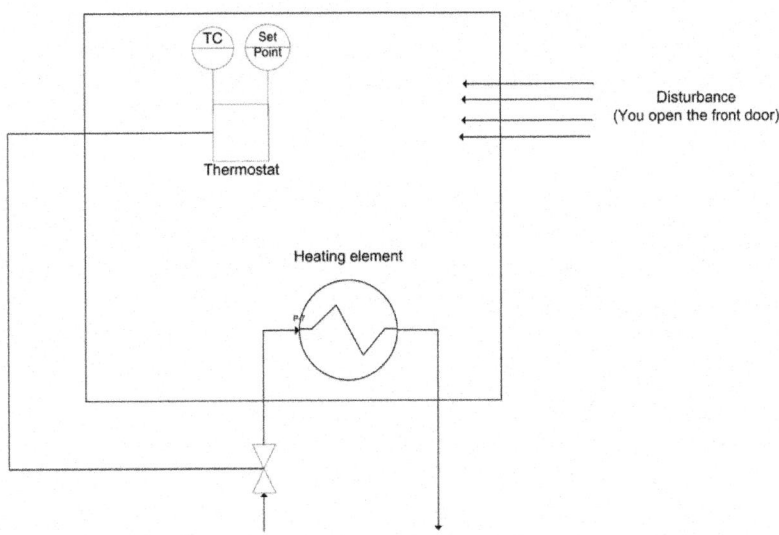

Closedloop System
(Winter Household Heating)

In this diagram a thermostat presents feedback to the heater to turn it on or off.

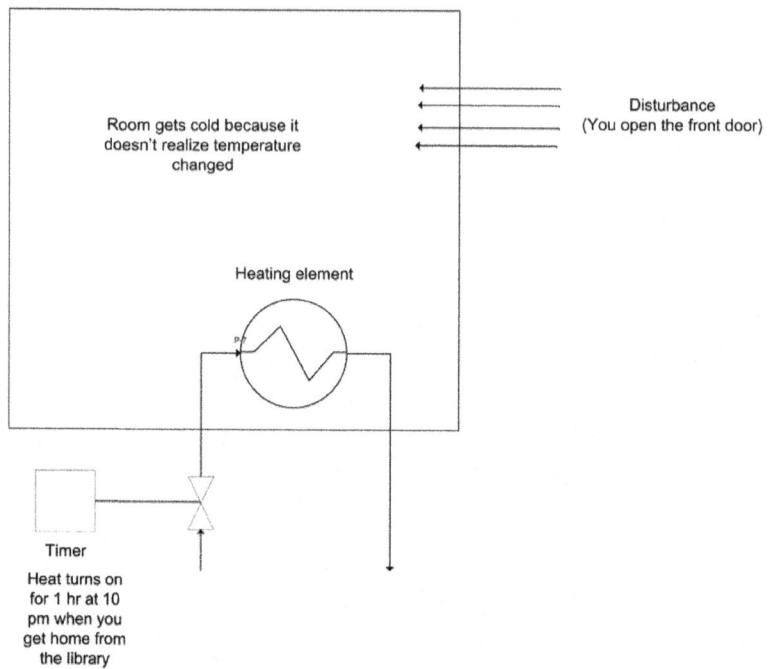

Openloop System
(Winter Household Heating)

Since the heater only turns on at 10 pm, it assumes that the amount of time you open the door will be the same each day. It is considered an open system since the temperature in the room is independent of the heater controller.

Note that all the examples presented in this web page discuss closed-loop systems because they include all systems with feedback.

Complex Coordinate Systems

Root locus plots are a plot of the roots of a characteristic equation on a complex coordinate system. A complex coordinate system allows the plotting of a complex number with both real and imaginary parts. The real component is plotted on the x-axis and the imaginary component is plotted on the y-axis. When creating root locus plots imaginary roots must be solved for. These imaginary roots come in complex conjugate pairs (this can be seen below in the section on "Plotting Poles on a Complex Coordinate System to make Root Locus Plot").

For example a plot of the following complex numbers is shown below. These complex numbers can be broken into the real and imaginary components to make it easier to plot.

Complex Number	Real Component	Imaginary Component
-0.4 + 0.2i	-0.4	0.2
-0.6 − 0.2i	-0.6	-0.2
-0.2 + 0i	-0.2	0
0 − 0.4i	0	-0.4
0.2 + 0.4i	0.2	0.4
0.4 − 0.6i	0.4	-0.6

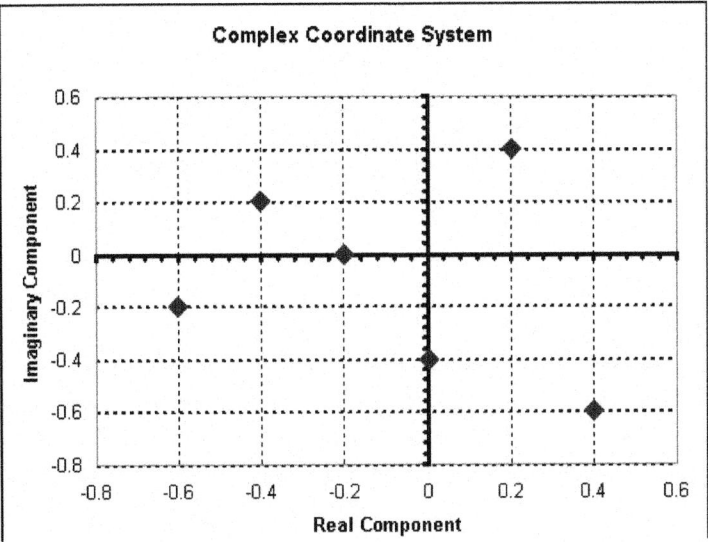

Developing a Characteristic Equation

Although the focus of this chapter is to discuss root locus plots, it is necessary to mention briefly how to determine the characteristic equation for a system in order to obtain the root locus plot. In general, most chemical engineering processes can be described by a system of ordinary differential equations. Follow the following steps to determine the characteristic equation for the system (which will allow you to develop a root locus plot).

1. If the ODEs are not linear, linearize them.

2. After linearizing the ODEs, use matrix algebra to find the eigenvalues of your system. Be careful here not to insert numerical values for your control parameters, (*e.g.* leave Kc as Kc, not Kc=1).

3. The polynomial equation obtained for the eigenvalue should contain lambda and the control parameters. This equation is the characteristic equation. Obtain solutions to this equation by setting values for the control parameters and solving for the eigenvalues. The roots obtained will be used to create the root locus diagram.

This 3-step process is valid to obtain a characteristic equation for any closed loop control system.

A more traditional method to develop characteristic equations is by applying Laplace transforms.

Laplace Transforms

Laplace transforms are a method to change linear ordinary differential equations into transfer function. All transfer functions used in root locus plots are independent of time because the $L[f(t)] \equiv F(s)$. Formally, the equation below shows that the time function is integrated, leaving only the variable s.

$$F(s) = \int_0^\infty f(t)e^{-st}dt$$

s is a complex number, therefore allowing us to construct complex coordinate system graphs. The exact solution to most disturbances and controllers can be found in any controls book.

Example

The stability of the series chemical reactors is to be determined. The reactors are well mixed and isothermal, and the reaction is first-order in component A. The outlet concentration of the second reactor is controlled with a PI feedback algorithm that manipulates the flow of the reactant, which is very much smaller than the flow of the solvent. The sensor and final element are assumed fast, and process data is as follows.

Process

$V = 5m^3$

$Fs = 5m^3/min >> FA$

$k = 1 \ min^{-1}$

$vs = 50\%$ open

$CA_0 = 20 \ mole/m^3$

$CA_0(s)/v(s) = Kv = 0.40(mole/ \ m^3)(\%open)$

PI Controller $Kc = ??Tf = 1 \ min$

Formulation The transfer function for the process and controller are

$$G_p(s) = \frac{K_p}{(\tau s + 1)(\tau s + 1)}$$

$$G_c(s) = K_c \left(1 + \frac{1}{T_I s} \right)$$

$$K_p = K_v \left(\frac{F}{F + Vk} \right) = 0.1 \frac{mole/m^3}{\%}$$

$$\tau = \frac{V}{F + Vk} = 0.5 \ min$$

The individual transfer functions can be combined to give the closed-loop transfer function for a set point change, which includes the characteristic equation. (where CV= Control variable & SP = set-point signal)

$$\frac{CV(s)}{SP(s)} = \frac{G_{p}(s)G_{v}(s)G_{c}(s)}{1+G_{p}(s)G_{v}(s)G_{c}(s)G_{s}(s)} = \frac{K_{p}(1+\frac{1}{s})\frac{0.1}{(0.5s+1)^{2}}}{1+K_{p}(1+\frac{1}{s})\frac{0.1}{(0.5s+1)^{2}}}$$

Characteristic equation

$$0 = 1+K_{p}(1+\frac{1}{s})\frac{0.1}{(0.5s+1)^{2}}$$

Root Locus Diagrams

Root Locus plots are a method of evaluating the behaviour of a control system. The creation of a root locus plot begins by determining the poles of the control system for a given set of control parameters. These poles are then plotted on a complex coordinate system as seen in the previous section and analyzed to determine the behaviour of the system.

Determining the Poles of a Control System

The "poles" of a system are the roots of the demoninator of the transfer function. In other words the poles are the values of "s" when the transfer function go to infinity (when the demoninator equals zero). For a system of differential equations, poles are the eigenvalues of the equation system. Consider the following solution to a system of differential equations:

$$f(s) = 48s^3 + 44s^2 + 12s + 1$$

This equation is a third order polynomial, therefore it will have three poles (be aware that some of these poles may be imaginary numbers). These poles can be obtained by factoring the expression or using a computer program such as Maple. The three poles, or roots, of this equation are s = -0.167, -0.25, -0.5.

For a system of differential equations, finding the eigenvalues can be time consuming and the use of Matlab, Maple, or Mathematica is more efficient. One method would be to use the Mathematica **eigenvalues[]** function to solve the system for you.

Plotting Poles on a Complex Coordinate System to make Root Locus Plot

A root locus plot is created by plotting the resulting poles on a complex coordinate system. For system of P-only control, the governing differential equations will depend on the proportional gain, Kc. Consider the following solution to a system of differential equations:

$$f(s) = 48s^3 + 44s^2 + 12s + 1 + 6Kc$$

Notice that Kc is a term in this equation. Therefore, there exists a set of poles for each value of Kc. If Kc = 0, then equation #2 reduces to equation #1, and poles are as listed above. Table lists the three poles of the system for given values of Kc. These poles were calculated using a computer algebra system because it is impractical to try and evaluate these functions with analytical methods.

Table. Poles of Characteristic Equation.

Kc	Pole 1	Pole 2	Pole 3
-0.028	-0.406	-0.406	-0.105
0.000	-0.167	-0.25	-0.500
0.004	-0.205	-0.205	-0.506
0.050	-.182+.126i	-.182-.126i	-0.553
0.200	-.140+.229i	-.140-.229i	-0.637
0.500	-.093+.325i	-.093-.325i	-0.731
1.000	-.045+.417i	-.045-.417i	-0.880
1.670	0-.500i	0+.500i	-0.917
2.000	.018-.532i	.018+.532i	-0.952

To create a root locus plot, each pole is broken down into its real (x-axis) and imaginary (y-axis) component:

Table. Real and Imaginary Components of each Pole.

Kc	Real Part	Imaginary Part
0	-0.167	0
0	-0.25	0
0	-0.5	0
0.004	-0.205	0
0.004	-0.205	0
0.004	-0.506	0
0.05	-0.182	0.126
0.05	-0.182	-0.126
0.05	-0.553	0
0.2	-0.14	0.229
0.2	-0.14	-0.229
0.2	-0.637	0
0.5	-0.093	0.325
0.5	-0.093	-0.325
0.5	-0.731	0
1	-0.045	0.417
1	-0.045	-0.417
1	-0.88	0
1.67	0	0.5
1.67	0	-0.5
1.67	-0.917	0
2	0.018	0.532
2	0.018	-0.532
2	-0.952	0

These points can now be plotted to make the root-locus diagram:

Root Locus Plot

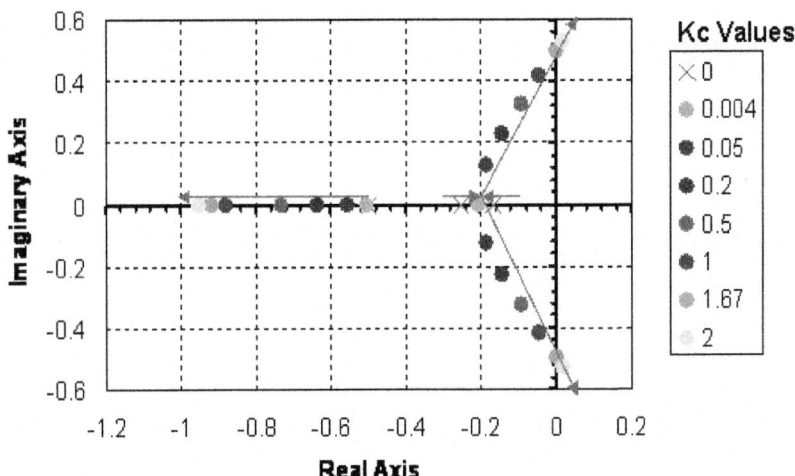

Fig. : Root Locus Diagram of the Characteristic Equation.

By convention, red arrows are drawn on the plot in the direction of increasing Kc values. They help to illustrate how the roots of the system vary by changing the Kc value. Also by convention, the points in which Kc = 0 are represented with 'x' marks instead of dots.

Interpreting a Root Locus Diagram

The primary use of a Root Locus Diagram is to evaluate how differing values of Kc affect the stability and behaviour of a control system.

The stability of the control system depends on the sign of the real component of the pole. If the real components of all poles are negative, then the system is said to be stable for that value of Kc. If the real component of the pole is positive, the system is unstable for that value of Kc, meaning the output signal will diverge from the set point.

The behaviour of the control system depends on the presence of an imaginary component of the pole. If any of the three poles contains an imaginary number component, then that value of Kc will cause the output signal to oscillate. If all of the poles are real (contain no imaginary components), the output signal will not oscillate at that Kc value.

Please refer back to the Root Locus diagram in Figure 1. The system becomes unstable (*i.e.* the real component of the poles becomes positive) for Kc>1.67 and Kc<0. Within the range of stability, no oscillations (*i.e.* no imaginary component) are observed when 0<Kc<0.004.

Root Locus Diagrams for PID Control

Root Locus diagrams are much more difficult to create for PID control. The characteristic equation will contain unknown variables Kc, Ti, and Td. Therefore, each point on the Root Locus Diagram will represent a set of tuning parameters. In order to show the progression as each tuning parameter changes, the resulting diagram will be a three-dimensional surface plot. Due to the complexity of this diagram, we will not create one, as it is beyond the scope of this text. However, stability analysis can still be applied to the characteristic equation. Take for example equation 3:

$$f(s) = \frac{48s^3 + 44s^2 + 12s + 1 + 6Kc}{Ti + 3Td}$$

In this theoretical instance, the given equation is the characteristic equation governing the same system mentioned above, only now with PID control. Notice the additional presence of the Ti and Td terms. Under P- only control, this system was stable for Kc values between 0 and 1.67. Will the same be true if this system were tuned with PID control?

Let's suppose we want to test the conditions Kc = 1.0, Ti = 0.3 and Td = 0.1. These values were randomly chosen. Using a computer software package, the three roots were determined to be: r1 = -0.827, r2 = -0.044 + 0.417i, r3 = -0.044 - 0.417i. Therefore, since the real components are all negative, the system is still stable. Since complex roots are present, the response is expected to oscillate around the set point.

Creating Root Locus Plots with Mathematica

Mathematica allows you to develop root locus plots for polynomials since the math involved in solving for the solutions can become very tedious. Before you are able to obtain the root locus plot, you need to solve for the roots. We will use the same equation as was used in the "Determining the Poles of a Control System" section.

$$f(s) = 48s^3 + 44s^2 + 12s + 1 + 6Kc$$

The **Solve[]** function can be used to determine the roots, both real and imaginary, for each corresponding Kc value. The syntax in Mathematica is shown below for the specified equation when Kc = 0. It is important to note that in order to obtain an output you must press "Shift Enter."

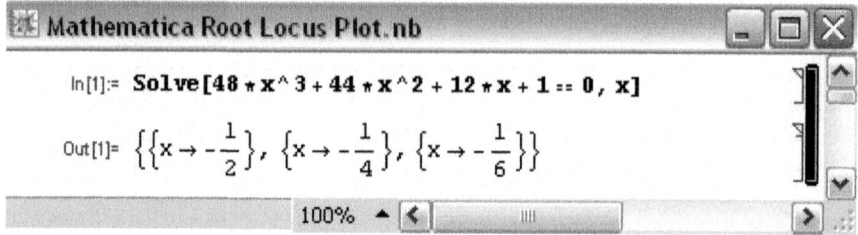

The following is just to show how Mathematica formats the output when Kc ≠ 0 and when there are imaginary roots. In this case, Kc = -0.167.

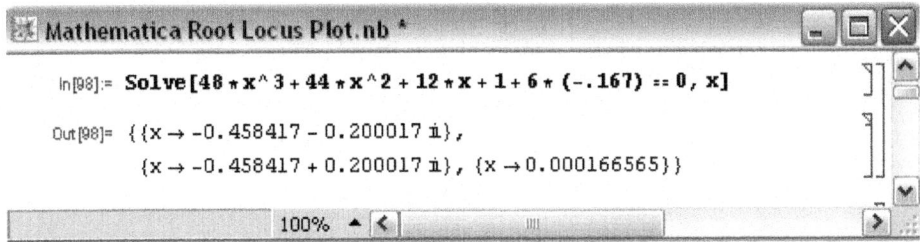

In[98]:= **Solve[48 * x ^ 3 + 44 * x ^ 2 + 12 * x + 1 + 6 * (-.167) == 0, x]**

Out[98]= {{x → -0.458417 - 0.200017 i},
{x → -0.458417 + 0.200017 i}, {x → 0.000166565}}

This can be done for all Kc values to obtain the corresponding real and imaginary roots. After all of the roots have been calculated, a table can be made to format the real roots (x axis) verses imaginary roots (y axis). When the syntax, **A={{x,y}{a,b}...}**, is used, you are inputting all of the x and y values and naming those values **A**. When you call **A** in the **TableForm[]** function, the table will be made according to all of the x and y values input into **A**. The syntax and corresponding outputs are shown below.

A = {{-0.466, 0.217}, {-0.466, -0.217}, {0.016, 0}, {-0.458, 0.2},
{-0.458, -0.2}, {0, 0}, {-0.406, 0}, {-0.406, 0}, {-0.105, 0},
{-0.167, 0}, {-0.25, 0}, {-0.5, 0}, {-0.205, 0}, {-0.205, 0},
{-0.506, 0}, {-0.182, 0.126}, {-0.182, -0.126}, {-0.533, 0},
{-0.14, 0.229}, {-0.14, -0.229}, {-0.637, 0}, {-0.093, 0.325},
{-0.093, -0.325}, {-0.731, 0}, {-0.045, 0.417}, {-0.045, -0.417},
{-0.88, 0}, {0, 0.5}, {0, -0.5}, {-0.917, 0}, {0.018, 0.532},
{0.018, -0.532}, {-0.952, 0}}

Out[104]= {{-0.466, 0.217}, {-0.466, -0.217}, {0.016, 0}, {-0.458, 0.2},
{-0.458, -0.2}, {0, 0}, {-0.406, 0}, {-0.406, 0},
{-0.105, 0}, {-0.167, 0}, {-0.25, 0}, {-0.5, 0}, {-0.205, 0},
{-0.205, 0}, {-0.506, 0}, {-0.182, 0.126}, {-0.182, -0.126},
{-0.533, 0}, {-0.14, 0.229}, {-0.14, -0.229}, {-0.637, 0},
{-0.093, 0.325}, {-0.093, -0.325}, {-0.731, 0}, {-0.045, 0.417},
{-0.045, -0.417}, {-0.88, 0}, {0, 0.5}, {0, -0.5},
{-0.917, 0}, {0.018, 0.532}, {0.018, -0.532}, {-0.952, 0}}

```
Mathematica Root Locus Plot cont.nb *                                    _ □ X
```

In[107]:=

```
TableForm[A, TableAlignments → Right,
    TableHeadings -> {Automatic, {"Real", "Imaginary"}}]
```

Out[107]//TableForm=

	Real	Imaginary
1	-0.466	0.217
2	-0.466	-0.217
3	0.016	0
4	-0.458	0.2
5	-0.458	-0.2
6	0	0
7	-0.406	0
8	-0.406	0
9	-0.105	0
10	-0.167	0
11	-0.25	0
12	-0.5	0
13	-0.205	0
14	-0.205	0
15	-0.506	0
16	-0.182	0.126
17	-0.182	-0.126
18	-0.533	0
19	-0.14	0.229
20	-0.14	-0.229
21	-0.637	0
22	-0.093	0.325
23	-0.093	-0.325
24	-0.731	0
25	-0.045	0.417
26	-0.045	-0.417
27	-0.88	0
28	0	0.5
29	0	-0.5
30	-0.917	0
31	0.018	0.532
32	0.018	-0.532
33	-0.952	0

```
                              100%  ▲ ◀ |          IIII          | ▶
```

In order to produce the table shown above, you only need **TableForm[A]** and nothing else. The other descriptions (*i.e.* TableAlignments and TableHeadings) are just used for formatting purposes.

Once the real (x) and imaginary (y) roots have been determined and put into table format for your equation, the **ListPlot[]** function can be used to develop your root locus diagram by calling **A**.

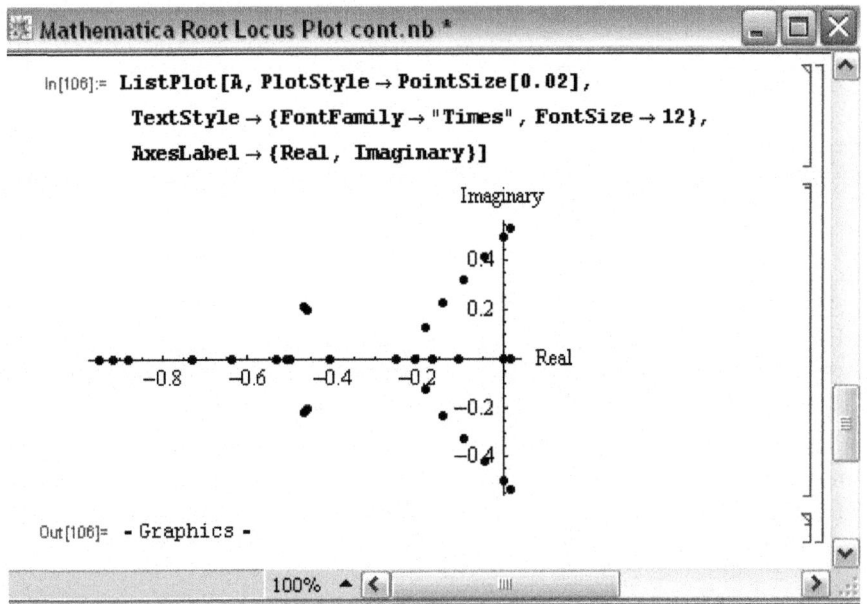

As you can see, the **ListPlot[]** function returned the same plot as shown in the "Complex Coordinate System to make Root Locus Plot" section (this plot was created in Excel). The arrows are not shown in this Mathematica plot; however the Excel plot shows the direction of increasing Kc values to show how the roots of the system vary by changing the Kc values.

The Mathematica Help Browser is a very useful tool for understanding the syntax. For example, when using the **ListPlot[]** function you can search how to label the axes and how to make the points larger (more visible) on the plot. The following examples are used to illustrate the use of root locus plots.

Second Plot Method Using Arrays

The goal here is to be able to understand exactly what is happening with these Mathematica inputs in order to get a better grasp on the outputs. Basically, once you have found the Eigenvalues of the system, which examples can be found for in other areas of the wiki, we then want to solve explicitly for them. After we solve for the Eigenvalue in terms of the variables, we can create an array by varying this value and solving for the real and imaginary terms of the Eigenvalues.

Eigenvalue Outputs in Mathematica

The Eigenvalues can either be returned as a function of 'l', which is the designation of the Eigenvalue or simply as a complex function in terms of variables. When returned as a function of 'l' there will be a polynomial input that has to be simplified to find 'l' as a function of the variable. In this case: Kc.

```
Eigenvalues[Jac /. param]
```

$$\left\{\left\{-\frac{1}{2}, \ 0, \ -\frac{Kc}{2}\right\}, \ \left\{\frac{1}{4}, \ -\frac{1}{4}, \ 0\right\}, \ \left\{0, \ \frac{1}{6}, \ -\frac{1}{6}\right\}\right\}$$

$$\left\{\text{Root}\left[1 + Kc + 12\,\sharp 1 + 44\,\sharp 1^2 + 48\,\sharp 1^3 \ \&, \ 1\right],\right.$$
$$\text{Root}\left[1 + Kc + 12\,\sharp 1 + 44\,\sharp 1^2 + 48\,\sharp 1^3 \ \&, \ 2\right],$$
$$\left.\text{Root}\left[1 + Kc + 12\,\sharp 1 + 44\,\sharp 1^2 + 48\,\sharp 1^3 \ \&, \ 3\right]\right\}$$

```
Kc =.
ss = N[Simplify[Solve[48*1^3 + 44*1^2 + 12*1 + (1 + Kc) == 0, 1]]];
```

What is happening here is that the Jacobian was inputted, and there were multiple eigenvalue arrays that were found. The last array has three complex polynomial functions that equal to 0. The last line is simplifying them, solving for 'l'. This is putting the simplified Eigenvalues into the array 'ss'.

Another output of Eigenvalues is simply in the form of complex numbers themselves. Rather than worry about the simplification steps, these Eigenvalues can be directly inputted as a complex number into an array, again termed 'ss':

$$ss = \left\{\left\{1 \rightarrow \left(-5 - 10*Kc - \sqrt{150 - 1590*Kc + 1240*Kc}\right)\right\},\right.$$
$$\left.\left\{1 \rightarrow \left(-5 - 10*Kc + \sqrt{150 - 1590*Kc + 1240*Kc}\right)\right\}\right\}$$

Plot Array Creation

Now that we have seen the two ways of Eigenvalue array creation, we can use these to create another array that will be the points for the plot. Here we will be substituting in values of the variable Kc to find different values of the Eigenvalues, which will give us information on stability changes as Kc changes.

```
Kc =.
ss = N[Simplify[Solve[48*1^3 + 44*1^2 + 12*1 + (1 + Kc) == 0, 1]]];

a = {};
Kc = 0;
a = Append[a, {Re[1 /. ss[[2]]], Im[1 /. ss[[2]]]}];
a = Append[a, {Re[1 /. ss[[1]]], Im[1 /. ss[[1]]]}];
a = Append[a, {Re[1 /. ss[[3]]], Im[1 /. ss[[3]]]}];
Kc = 0.5;
a = Append[a, {Re[1 /. ss[[2]]], Im[1 /. ss[[2]]]}];
a = Append[a, {Re[1 /. ss[[1]]], Im[1 /. ss[[1]]]}];
a = Append[a, {Re[1 /. ss[[3]]], Im[1 /. ss[[3]]]}];
```

In the example above, 'ss' is in the form : {{solution1}, {solution2}, {solution3}} Each solution in this array is in the form l ->a+bi. For example, ss[[2]] refers to solution2, which is the object in the second position in ss. Re[l/.ss[[2]]] applies solution2 to the variable l and then returns only the real part. Im[l/.ss[[2]]] applies solution2 to the variable l and then returns only the imaginary part. The effect of the Append function above is that the ordered pair {a,b} (from a solution of the form a+bi) is added to the array called 'a'.

However, not all Eigenvalue arrays will have the same number of Eigenvalues. For example, as mentioned in the above paragraph, 'ss' is in the form : {{solution1}, {solution2}, {solution3}} There could be 2, or 3, or more separate 'solution' values in the array. Depending on the number of the 'solution' values, there will be a different format for your 'append' function in Mathematica. For example, in the image below, there are only two eigenvalues in our arrat that we want to solve for. Thus, there should only be two rows of the addition to the array for the this method:

```
a = {};
F = 0;
a = Append[a, {Re[1 /. ss[[2]]], Im[1 /. ss[[2]]]}];
a = Append[a, {Re[1 /. ss[[1]]], Im[1 /. ss[[1]]]}];
```

These different methods both output arrays of numbers that act essentially as 'x' and 'y' values, or in this case Real and Imaginary, for the plot. The array 'a' is plotted as such:

ListPlot[a, PlotStyle ->PointSize[0.02]]

This is inputted directly to Mathematica, which creates the Root Locus Plot.

Differential Equation Example of Root Locus Plots in Mathematica

An example problem pulled from the Fall2008 second exam:

"You have just been put in charge of designing a large-scale bioreactor for the production of a blockbuster protein based drug. After some research, you have come up with the following model to describe the reactor system:"

$dX/dt = -2X + (10 * X * S)/(S + 2)$

$dS/dt = 4 * Fin - S - (3 * X * S)/(S + 2)$

$dP/dt = 9 * X - 2 * P$

X is the cell concentration, S is the nutrient concentration, and P is the protein product. Fin is the flow rate of nutrient into the system.

In this system there are two fixed points found by setting all the differential equations equal to zero and solving in Mathematica.

```
eqn = {0 == -2*X + (10*X*S) / (S+2), 0 == 4*Fin - S - (3*X*S) / (S+2), 0 == 9*X - 2*P};
eqns = {-2*X + (10*X*S) / (S+2), 4*Fin - S - (3*X*S) / (S+2), 9*X - 2*P};
Solve[eqn, {X, S, P}]
```

Out[4]= $\left\{ \{P \to 0, X \to 0, S \to 4\,Fin\}, \left\{ P \to \frac{15}{4}\,(-1 + 8\,Fin), X \to \frac{5}{6}\,(-1 + 8\,Fin), S \to \frac{1}{2} \right\} \right\}$

The two fixed points depend on the value of Fin and with that varying values of Fin will change eventual eigenvalues. Since this problem will eventually want root locus plots for both fixed points it is easier to save both points as parameters to be applied to general equations later. Overall it will help save the need for redundant code.

From this point a Jacobian matrix should be created to linearize the system around the fixed points (as applied to the Jacobian). Once created the stored fixed point values for X, S, and P can be applied. Once applied, the eigenvalues for each fixed point can be solved for. These will still contain a variable 'Fin' since Fin has not been defined yet.

```
In[192]:= Clear[Fin]
        eqn = {0 == -2*X + (10*X*S) / (S+2), 0 == 4*Fin-S - (3*X*S) / (S+2), 0 == 9*X-2*P};
        eqns = {-2*X + (10*X*S) / (S+2), 4*Fin-S - (3*X*S) / (S+2), 9*X-2*P};
        Solve[eqn, {X, S, P}]
        par1 = {P → 0, X → 0, S → 4*Fin};
        par2 = {P → 15/4 (-1+8 Fin), X → 5/6 (-1+8 Fin), S → 1/2};
        Jac = {{D[eqns[[1]], X], D[eqns[[1]], S], D[eqns[[1]], P]},
               {D[eqns[[2]], X], D[eqns[[2]], S], D[eqns[[2]], P]},
               {D[eqns[[3]], X], D[eqns[[3]], S], D[eqns[[3]], P]}};
        a = Jac /. par1;
        b = Jac /. par2;
        MatrixForm[a]
        MatrixForm[b]
        c = N[Eigenvalues[a]];
        d = N[Eigenvalues[b]];
```

Out[201]//MatrixForm=

$$
\begin{pmatrix}
-2 + \dfrac{40\,Fin}{2+4\,Fin} & 0 & 0 \\[2mm]
-\dfrac{12\,Fin}{2+4\,Fin} & -1 & 0 \\[2mm]
9 & 0 & -2
\end{pmatrix}
$$

Out[202]//MatrixForm=

$$
\begin{pmatrix}
0 & \dfrac{8}{3}(-1+8\,Fin) & 0 \\[2mm]
-\dfrac{3}{5} & -8\,Fin + \dfrac{1}{5}(-1+8\,Fin) & 0 \\[2mm]
9 & 0 & -2
\end{pmatrix}
$$

From here the Table[] function can be used to tabulate eigenvalues for various Fin values for visual sake of inspection. This is not necessary since the eigenvalue results have already been solved for in terms of Fin. To create the root locus plots, values of Fin need to be defined and applied to the eigenvalue equations with the results stored. That step is next. In this table step it is for the user's sake to see with increasing Fin values what the eigenvalues look like. The tabulated eigenvalues are stored as a matrix and displayed as such with the function 'MatrixForm[]'

```
a = Jac /. par1;
b = Jac /. par2;
MatrixForm[a]
MatrixForm[b]
c = N[Eigenvalues[a]];
d = N[Eigenvalues[b]];
e = Table[c, {Fin, 0, 2, 0.1}];
MatrixForm[e]
f = Table[d, {Fin, 0, 2, 0.1}];
MatrixForm[f]
```

Out[206]//MatrixForm=

$$
\begin{pmatrix}
-2. & -1. & -2. \\
-2. & -1. & -0.333333 \\
-2. & -1. & 0.857143 \\
-2. & -1. & 1.75 \\
-2. & -1. & 2.44444 \\
-2. & -1. & 3. \\
-2. & -1. & 3.45455 \\
-2. & -1. & 3.83333 \\
-2. & -1. & 4.15385 \\
-2. & -1. & 4.42857 \\
-2. & -1. & 4.66667 \\
-2. & -1. & 4.875 \\
-2. & -1. & 5.05882 \\
-2. & -1. & 5.22222 \\
-2. & -1. & 5.36842 \\
-2. & -1. & 5.5 \\
-2. & -1. & 5.61905 \\
-2. & -1. & 5.72727 \\
-2. & -1. & 5.82609 \\
-2. & -1. & 5.91667 \\
-2. & -1. & 6.
\end{pmatrix}
$$

Out[208]//MatrixForm=

$$
\begin{pmatrix}
-2. & -1.36886 & 1.16886 \\
-2. & -1.12456 & 0.284557 \\
-2. & -0.74 - 0.642184\,i & -0.74 + 0.642184\,i \\
-2. & -1.06 - 1.0566\,i & -1.06 + 1.0566\,i \\
-2. & -1.38 - 1.27106\,i & -1.38 + 1.27106\,i \\
-2. & -1.7 - 1.38203\,i & -1.7 + 1.38203\,i \\
-2. & -2.02 - 1.41407\,i & -2.02 + 1.41407\,i \\
-2. & -2.34 - 1.37273\,i & -2.34 + 1.37273\,i \\
-2. & -2.66 - 1.25076\,i & -2.66 + 1.25076\,i \\
-2. & -2.98 - 1.01961\,i & -2.98 + 1.01961\,i \\
-2. & -3.3 - 0.556776\,i & -3.3 + 0.556776\,i \\
-2. & -4.41019 & -2.82981 \\
-2. & -5.26801 & -2.61199 \\
-2. & -6.02284 & -2.49716 \\
-2. & -6.73787 & -2.42213 \\
-2. & -7.4318 & -2.3682 \\
-2. & -8.11282 & -2.32718 \\
-2. & -8.78524 & -2.29476 \\
-2. & -9.4516 & -2.2684 \\
-2. & -10.1135 & -2.2465 \\
-2. & -10.772 & -2.228
\end{pmatrix}
$$

These tables make it easy to see with increasing Fin (goes from 0 to 2 down the table) what the changes are in the stability of each fixed point. The first table is the first fixed point, the second table the second FP.

Now, to make the root locus plots, a really long string of values needs to be created in order for the real and imaginary roots to be plotted. This is a long string of code that is basically brute force. The overall action occurring here is setting Fin to a value, evaluating that Fin through the eigenvalues, and saving the real portion separate from the imaginary portion by different columns. This is repeated for as many Fin values as you desire. It is extremely repetitive, but works. After evaluating as many Fin values as you desired you can display the two columns through the ListPlot[] function. Example snippets of the code look like such:

```
Fin =.
rootc = {}
Fin = 0;
rootc = Append[rootc, {Re[c[[2]]], Im[c[[2]]]}];
rootc = Append[rootc, {Re[c[[1]]], Im[c[[1]]]}];
rootc = Append[rootc, {Re[c[[3]]], Im[c[[3]]]}];
Fin = 0.1;
rootc = Append[rootc, {Re[c[[2]]], Im[c[[2]]]}];
rootc = Append[rootc, {Re[c[[1]]], Im[c[[1]]]}];
rootc = Append[rootc, {Re[c[[3]]], Im[c[[3]]]}];
Fin = 0.2;
rootc = Append[rootc, {Re[c[[2]]], Im[c[[2]]]}];
rootc = Append[rootc, {Re[c[[1]]], Im[c[[1]]]}];
rootc = Append[rootc, {Re[c[[3]]], Im[c[[3]]]}];
Fin = 0.3;
rootc = Append[rootc, {Re[c[[2]]], Im[c[[2]]]}];
rootc = Append[rootc, {Re[c[[1]]], Im[c[[1]]]}];
rootc = Append[rootc, {Re[c[[3]]], Im[c[[3]]]}];
Fin = 0.4;
rootc = Append[rootc, {Re[c[[2]]], Im[c[[2]]]}];
rootc = Append[rootc, {Re[c[[1]]], Im[c[[1]]]}];
rootc = Append[rootc, {Re[c[[3]]], Im[c[[3]]]}];
Fin = 0.5;
rootc = Append[rootc, {Re[c[[2]]], Im[c[[2]]]}];
rootc = Append[rootc, {Re[c[[1]]], Im[c[[1]]]}];
rootc = Append[rootc, {Re[c[[3]]], Im[c[[3]]]}];
```

And the ListPlot[] function can look like this:

```
ListPlot[rootc, PlotStyle → PointSize[0.02],
    TextStyle → {FontFamily → "Times", FontSize → 12}, AxesLabel → {Real, Imaginary}]
```

And would make graphs like these (Fixed Point 1 on top and FP2 on bottom in this example):

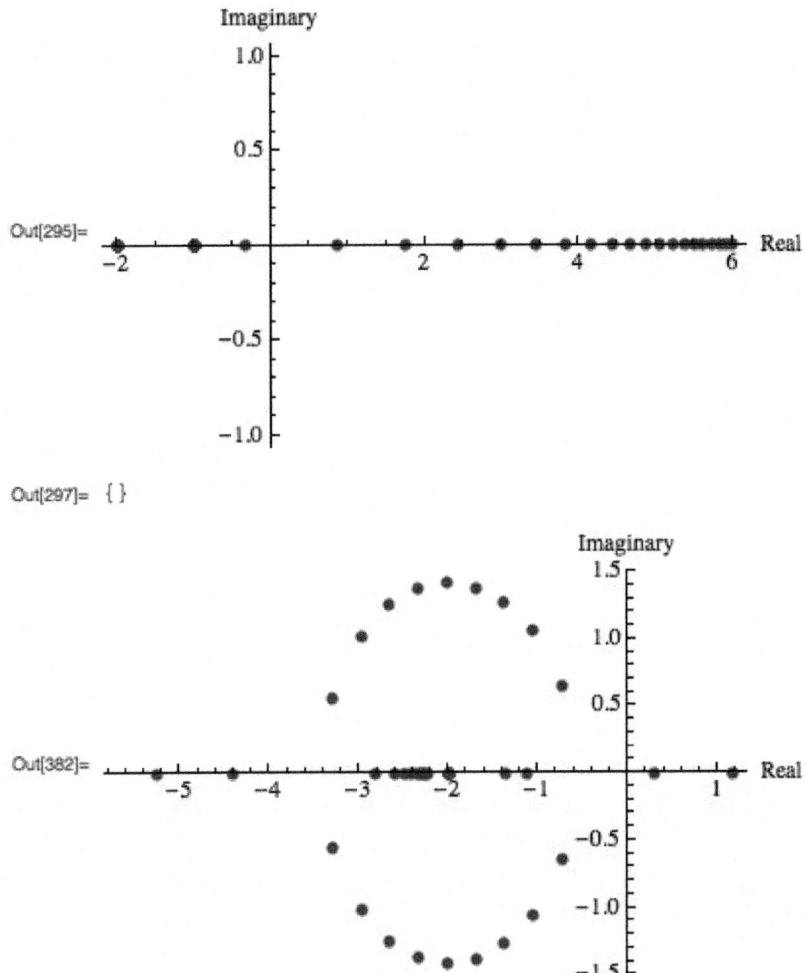

Looking back at the tables of eigenvalues the plots can be interpreted. For FP1 the values start real negative and increase to real positive. There are no imaginary values. For FP2, increasing Fin values go from real positive to negative with imaginary values occurring in the transition from low to high Fin values.

Alternative Mathematica Method

Some Mathematica6 programs have an add-on or supplementary application known as **ANALOG INSYDES** which contains special options to create Root Locus Plots with ease. The Function is called *RootLocusPlot*. Enter the function that is to be analyzed, into Mathematica with the following format:

RootLocusPlot[*tfunc*,{*k*,*k_0*,*k_1*}]

tfunc is the transfer function in the frequency variable *s* and one real parameter *k*. k_0 and k_1 are the range for the real parameter *k* that is to be varied int he Root Locus Plot. Follow an example below.

Example: Equation to make a Root Locus Plot with:

$H(s) = (a + 2 * s + s^2)/ (10 + 3 * a * s + 4 * s^2 + s^3)$

To make a Root Locus Plot follow the example code below:

H4[s_, a_] := (a + 2*s + s^2)/(10 + 3*a*s + 4*s^2 + s^3)

RootLocusPlot[H4[s, a], {a, 3, 5}]

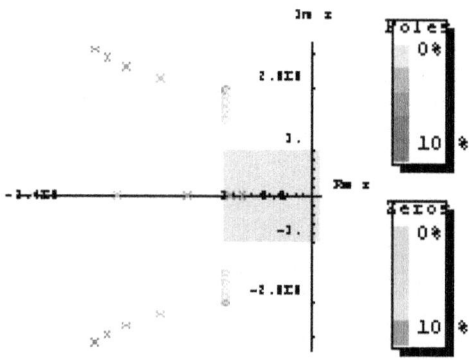

Image: Wolfram.com

Other Mathematica forms for using RootLocusPlot[]:

RootLocusPlot[*func*] This form displays the a pole/zero diagram of a function *i.e. func* without parameters *k* and *k_1*.

RootLocusPlot[*rootloc*] This form displays a root locus calculated with function RootLocusByQZ[].

Creating Root Locus Plots with Matlab

Root Locus Generation in Matlab Three matlab files have been given to obtain the root locus plot and poles of the root locus plot at specified Kc values for a specific transfer function with relative ease. The transfer function file is where the specific transfer function should be input. As written the only line needing variation is line 4(Gs). The poles function file finds the poles of the given transfer function using built in matlab utilities and needs no variation, even after changing the transfer function. The Locusplotpoles file is what is called in matlab to generate the root locus plot as well as the value of the poles and integrates the two previous files. To use this file you type in the values of the specific Kc you require the pole values for as well as a single value of Kc for which you would like to see the root locus plot. An example of the inputs to and outputs of this file are given below.

>>Locusplotpoles([050100200],0)

ans =

1.0e+002 *

Columns 1 through 3

0 -0.0020 -0.0020 + 0.0000i

0.5000 -0.0045 -0.0008 + 0.0022i

1.0000 -0.0051 -0.0004 + 0.0027i

2.0000 -0.0060 -0.0000 + 0.0034i

Columns 4 through 6

-0.0020 - 0.0000i 00

-0.0008 - 0.0022i   -0.0020 -0.0020 + 0.0000i

-0.0004 - 0.0027i   -0.0020 + 0.0000i -0.0020 - 0.0000i

-0.0000 - 0.0034i   -0.0020 -0.0020 + 0.0000i

Column 7

0

-0.0020 - 0.0000i

-0.0020

-0.0020 - 0.0000i

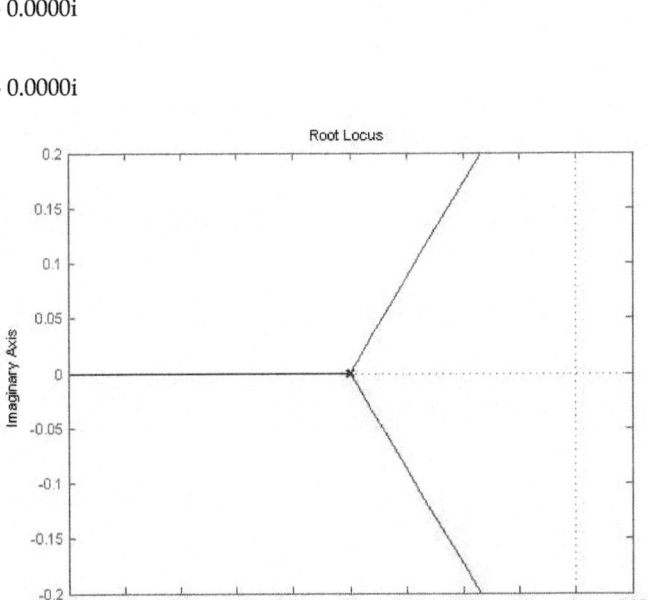

The values inside the brackets are the specific Kc values for which poles are desired. The last value is the Kc which the root locus plot will be plotted for. Notice that the value of Kc for the root locus plot to be generated in this example is zero. This is so that the entire range of Kc can be examined using the interactive graph produced by matlab. If other specific Kc plots would like to be observed this value can be changed to generate these as well. The column lines give the values of Kc input and then value of the poles for each Kc. Note that the actual values are the values matlab gives multiplied by 100.

This site gives animated examples which progress the value of the control variable through a root locus plot to better explain the fundamental function of these plots. It would be beneficial to go through a few of these examples to observe patterns of fluctuation of the control variable.

Creating Root Locus plots with Excel and PPLANE

Though this tactic may remain slightly more time consuming and work involved, there is a method in computing the Root Locus plot using PPLANE and Excel. This particular model really aids in better understanding of how to read and acknowledge a Root Locus Plot, and is also useful if alternative options for making a Root Locus Plot are not currently available.

This method involves using PPLANE in order to find the eigenvalues for each equilibrium value, and then plotting these points with Excel software. In order to do this, suppose that you have the following differential equations that represent a reactor:

$(dX/dt) = -X+((2+X+Y)/(Y+3))$

$(dY/dt) = Y*Fin+7-(2X)$

Graphing the following differential equations provides something similar to the following chart (a few lines were drawn, just to better show the graphs flow):

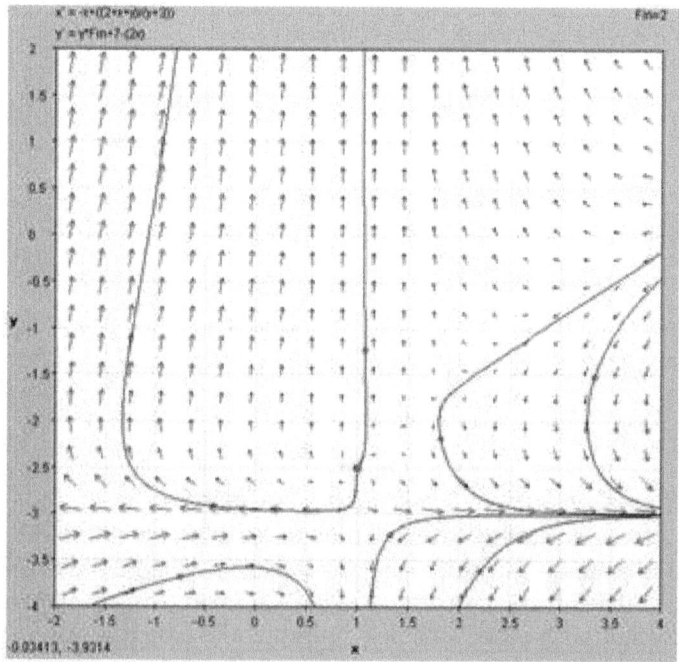

Using the ability to "Find an Equilibrium Point" in PPLANE, one can select a particular equilibrium point on the field, and be able to provide the following results:

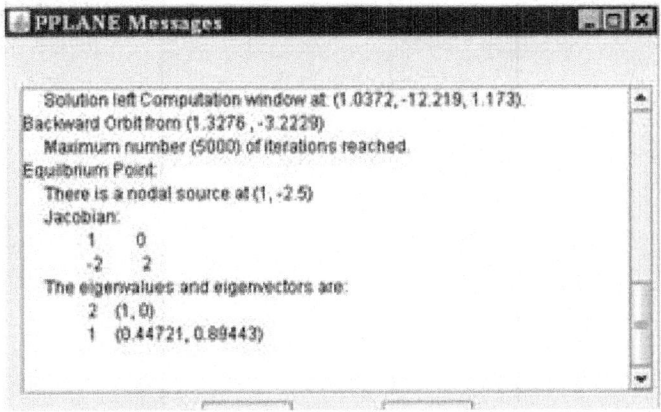

This window "pops up" in the higher left corner of the screen, when the equilibrium point is found. This window provides the eigenvalues for this equilibrium point.

Depending what you are varying (in this case, Fin), all you would have to do is enter various values of this Fin into PPLANE, and plot the new set of differential formulas. After this is done, just find the same equilibrium point, and record the new given eigenvalues in excel. These values should be entered with an x-component (in one column) and a y-component (in another column) into excel, with the following rules:

- Assuming the example 3+3i, 3-3i were given as eigenvalues.

- For any given eigenvalue, the x-value of a particular eigenvalue is the real part of that number (*i.e:* the "3" part).

- For any given point, the y-value of a particular eigenvalue is the imaginary part of that number (*i.e:* the "3i" part). If there is an imaginary component, ignore the "i," though the number should be recorded in the y-component. If there is no imaginary number, the y component is zero.

- For both cases, a positive is a positive, and a negative is a negative. Remember: imaginary numbers have 2 parts, a positive and negative "i" value.

These points are simply recorded according to these rules, then the points are plotted against each other, with x components on the x-axis, and the y in the y-axis. Here is an example of an excel diagram, taken from the proposed diagram:

X-axis	Y-axis (imaginary)
2.3028	0
-1.3028	0
2.2415	0
-2.1415	0
2.2471	0
-2.0471	0
4	0
-0.42857	0
2.3058	0
-0.85078	0
2.3252	0
-1.0752	0
2.4142	0
-0.41421	0
2.5	0
0	0

And, the excel data is then plotted, x against y, and is shown below:

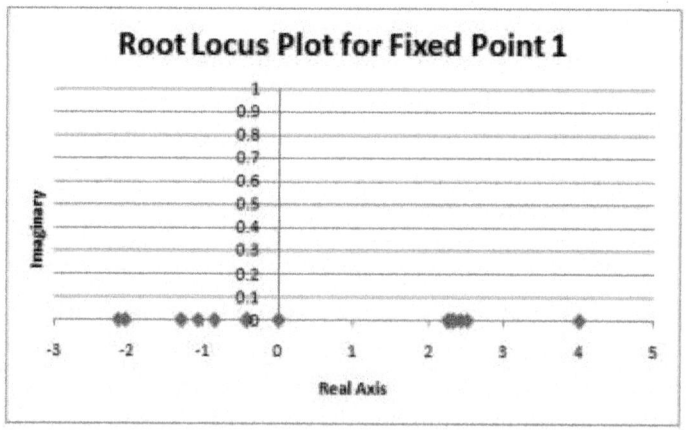

As you can see, this particular model did not have any imaginary data. The root locus plot has been formed, and for all accounts of Fin, the eigenvalues are along the x-axis for the root locus plot.

Practical Application

In the past, it was necessary for engineers to master the techniques required to efficiently construct root locus diagrams. In today's engineering world, this is not the case for one of two reasons. In many instances, root locus diagrams are not used industrially because they require models of the system which are generally not available. If a model is available to develop a root locus diagram, there are computer applications that can develop the diagrams much faster than a person. Thus, energy and effort should be placed on understanding and interpreting a root locus diagram and understanding the general rules of stability for a root of the characteristic equation.

ROUTH STABILITY

The stability of a process control system is extremely important to the overall control process. System stability serves as a key safety issue in most engineering processes. If a control system becomes unstable, it can lead to unsafe conditions. For example, instability in reaction processes or reactors can lead to runaway reactions, resulting in negative economic and environmental consequences.

The absolute stability of a control process can be defined by its response to an external disturbance to the system. The system may be considered stable if it exists at a consistent state or setpoint and returns to this state immediately after a system disturbance. In order to determine the stability of a system, one often must determine the eigenvalues of the matrix representing the system's governing set of differential equations. Unfortunately, sometimes the characteristic equation of the matrix (the polynomial representing its eigenvalues) can be difficult to solve; it may be too large or contain unknown variables. In this situation, a method developed by British mathematician Edward Routh can yield the desired stability information without explicitly solving the equation.

Recall that in order to determine system stability one need only know the signs of the real components of the eigenvalues. Because of this, a method that can reveal the signs without actual computation of the eigenvalues will often be adequate to determine system stability.

To quickly review, negative real eigenvalue components cause a system to return to a steady state point (where all partial derivatives equal zero) when it experiences disturbances. Positive real components cause the system to move away from a stable point, and a zero real component indicates the system will not adjust after a disturbance. Imaginary components simply indicate oscillation with a general trend in accordance with the real part. Using the method of Routh stability, one can determine the number of each type of root and thus see whether or not a system is stable. When unknown variables exist in the equation, Routh stability can reveal the boundaries on these variables that keep the roots negative.

The Routh Array

The Routh array is a shortcut to determine the stability of the system. The number of positive (unstable) roots can be determined without factoring out any complex polynomial.

Generating the Array

The system in question must have a characteristic equation of a polynomial nature as shown below:

$$P(S) = a_n S^n + a_{n-1} S^{n-1} + \cdots + a_1 S + a_0$$

In order to examine the roots, set P(S)=0, which will allow you to tell how many roots are in the left-hand plane, right hand plane, and on the j-omega axis. If the system involves trigonometric functions it needs to be fit to a polynomial via a Taylor series expansion. One necessary condition for stability is that $a_n >$ 0. (If $a_n < 0$, all coefficients may be multiplied by -1 before checking). The other condition is that all values in column 1 of the Routh array must be positive for the system to be stable.

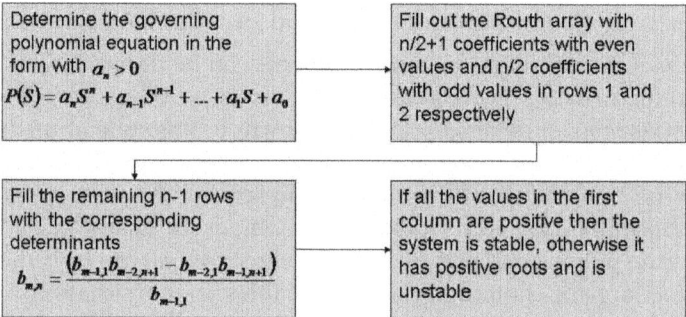

This flow diagram shows the generation of a Routh array for an idealized case with m,n representing the location in the matrix.

The coefficients of the polynomial are placed into an array as seen below. The number of rows is one more than the order of the equation. The number of sign changes in the first column indicate the number of positive roots for the equation.

Row 1	: a_n	a_{n-2}	a_{n-4}	$\cdots$
Row 2	: a_{n-1}	a_{n-3}	a_{n-5}	$\cdots$
Row 3	: b_1	b_2	b_3	$\cdots$
Row 4	: c_1	c_2	c_3	$\cdots$
:	: :	:	:	:
Row 5	: p_1	p_2		
Row 6	: q_1			
Row 7	: v_1			

In the array, the variables b1,b2,c1,c2,*etc.* are determined by calculating a determinant using elements from the previous two rows as shown below:

$$b_1 = \frac{a_{n-1}a_{n-2} - a_n a_{n-3}}{a_{n-1}}, b_2 = \frac{a_{n-1}a_{n-4} - a_n a_{n-5}}{a_{n-1}}, b_3 = \frac{a_{n-1}a_{n-6} - a_n a_{n-7}}{a_{n-1}}, \cdots$$

$$c_1 = \frac{b_1 a_{n-3} - a_{n-1}b_2}{b_1}, \quad c_2 = \frac{b_1 a_{n-5} - a_{n-1}b_3}{b_1}, c_3 = \frac{b_1 a_{n-7} - a_{n-1}b_4}{b_1}, \cdots$$

The general expression for any element x after the first two rows with index (m,n) is as follows:

$$\text{For } A = \begin{bmatrix} x_{m-2,1} & x_{m-2,n+1} \\ x_{m-1,1} & x_{m-1,n+1} \end{bmatrix}$$

$$x_{m,n} = \frac{-det(A)}{x_{m-1,1}}$$

Note, that if the Routh array starts with a zero, it may still be solved (assuming that all the other values in the row are not zero), by replacing the zero with a constant, and letting that constant equal a very small positive number. Subsequent rows within that column that have this constant will be calculated based on the constant choosen.

Once the array is complete, apply the following theorems to determine stability:

1. If all of the values in the fist column of the Routh array are >0, then P(S) has all negative real roots and the system is stable.

2. If some of the values in the first column of the Routh array are <0, then the number of times the sign changes down the first column will = the number of positive real roots in the P(S) equation.

3. If there is 1 pair of roots on the imaginary axis, both the same distance from the origin (meaning equidistant), then check to see if all the other roots are in the left hand plane. If so, then the imaginary roots location may be found using $AS^2 + B = 0$, where A and B are the elements in the Routh array for the 2nd to last row.

To clarify even further, an example with real numbers is analyzed.

Example Array

The following polynomial was generated from a sample system.

$$P(S) = 5S^3 - 10S^2 + 7S + 20$$

The preceding polynomial must be investigated in order to determine the stability of the system. This is done by generating a Routh array in the manner described above. The array as a result of this polynomial is,

Row 1	:	5	7
Row 2	:	-10	20
Row 3	:	17	
Row 4	:	20	

In the array shown above, the value found in the third row is calculated as follows.

$$b_1 = \frac{-10 * 7 - 5 * 20}{-10}$$

The array can now be analyzed. When looking down the first column, it can be seen that 5 is positive in magnitude, then the sign changes in the -10 entry, and the sign changes a second time to positive 17. This counts as two changes in sign, which corresponds to two positive roots, making the system unstable.

Finding Stable Control Parameter Values

Often, for a unit operation, a PID parameter such as controller gain (K_c), the integral time constant (T_i), or the derivative time constant (T_d) creates an additional variable in the characteristic equation. This can be carried through the computations of the Routh array to indicate which values of the variable will provide stability to the system through by preventing positive roots from occuring in the equation. For example, if a controller output is governed by the function:

$$10s^3 + 5s^2 + 8s + (T_d + 2)$$

The stable values of T_d can be found via a Routh array:

Row 1	:	10	8
Row 2	:	5	$(T_d + 2)$
Row 3	:	$4 - 2T_d$	
Row 4	:	$(T_d + 2)$	

We reveal $-2 < T_d < 2$ in order to keep the first column elements positive, so this is the stable range of values for this parameter. If multiple parameters were in the equation, they would simply be solved for as a group, yielding constraints along the lines of "$T_i + K_c > 2$" etc, so any value chosen for one parameter would give a different stable range for the other.

Special Cases

There are a few special cases that one should be aware of when using the Routh Test. These variances can arise during stability analysis of different control systems. When a special case is encountered, the traditonalRouth stability solution methods are altered as presented below.

One of the Coefficients in the Characteristic Equation Equals Zero

If the power of the $0*S^n$is ≥ 1, replace the zero with a quantity, ε, which would be positive and will approach zero. Then continue with your analysis as normal. Essentially this gives the limit of the roots as that coefficient approaches zero. (If the power = 0, see Case 3). For example,

Equation: $2S^3 - 24S + 32 = 0$

Working Equation: $2S^3 + \epsilon S^2 - 24S + 32 = 0$

Row 1	:	2	-24
Row 2	:	ϵ	32
Row 3	:	$\dfrac{-24\epsilon - 64}{\epsilon}$	0
Row 4	:	32	

Since ε is positive we know that in the first column row 2 will be positive, row 4 will be positive, and row 3 will be negative. This means we will have a sign change from 2 to 3 and again from 3 to 4. Because of this, we know that two roots will have positive real components. If you actually factor out the equation you see that

$$(S - 2)^2(2S + 8) = 0,$$

showing that we do have 2 positive roots. Both of these roots are equal to 2, so there is technically only one root, but in any case we know the system is unstable and must be redesigned.

One of the Roots is Zero

This case should be obvious simply from looking at the polynomial. The constant term will be missing, meaning the variable can be factored from every term. If you added an ε to the end as in case 1, the last row would be ε and falsely indicate another sign change. Carry out Routh analysis with the last zero in place.

Equation: $S^3 - S^2 - 2S = 0$

Row 1	:	1	-2
Row 2	:	-1	0
Row 3	:	-2	
Row 4	:	0	

As you can see in column one we have row 1 positive, row 2 and 3 negative, and row 4 zero. This is interpreted as one sign change, giving us one positive real root. Looking at this equation in factored form,

$$(S+1)(S-2)S = 0$$

we can see that indeed we have only one positive root equal which equals 2. The zero in the last row indicates an additional unstable root of zero. Alternatively, you may find it easier to just factor out the variable and find the signs of the remaining eigenvalues. Just remember there is an extra root of zero.

A Row Full of Zeros

When this happens you know you have either a pair of imaginary roots, or symmetric real roots. The row of zeros must be replaced. The following example illustrates this procedure.

Equation: $S^4 - 6S^3 + 10S^2 - 6S + 9 = 0$

Row 1	:	1	10	9
Row 2	:	-6	-6	
Row 3	:	9	9	
Row 4	:	0	0	

Row 4 contains all zeros. To determine its replacement values, we first write an auxiliary polynomial A determined by the entries in Row 3 above.

$$A(S) = 9S^2 + 9$$

Notice that the order decreases by 1 as we go down the table, but decreases by 2 as we go across.

We then take the derivative of this auxiliary polynomial.

$$A'(S) = 18S$$

The coefficients obtained after taking the derivative give us the values used to replace the zeros. From there, we can proceed the table calcuations normally. The new table is

Row 1	:	1	10	9
Row 2	:	-6	-6	
Row 3	:	9	9	
Row 4	:	18	0	
Row 5	:	9		

In fact, the purely imaginary or symmetric real roots of the original polynomial are the same as the roots of the auxiliary polynomial. Thus, we can find these roots.

$$9S^2 + 9 = 0$$

$$S = \pm i$$

Because we have two sign changes, we know the other two roots of the original polynomial are positive.

In fact, after factoring this polynomial, we obtain

$$(S^2 + 1)(S - 3)^2 = 0$$

Therefore, the roots are $S = \pm i, 3$, where in this case, the root 3 has multiplicity 2.

Limitations

Routh arrays are useful for classifying a system as stable or unstable based on the signs of its eigenvalues, and do not require complex computation. However, simply determing the stability is not usually sufficient for the design of process control systems. It is important to develop the extent of stability as well as how close the system is to instability. Further stability analysis not accounted for in the Routh analysis technique include finding the degree of stability, the steady state performance of the control system, and the transient response of the system to disturbances.

Note that for defining stability, we will always start out with a polynomial. This polynomial arises from finding the eigenvalues of the linearized model. Thus we will never encounter other functions, say exponenential functions or sin or cos functions in general for stability analysis in control theory.

Advantages Over Root Locus Plots

Routh stability evaluates the signs of the real parts of the roots of a polynomial without solving for the roots themselves. The system is stable if all real parts are negative. Therefore unlike root locus plots, the actual eigenvalues do not need to be calculated for a Routh stability analysis. Furthermore, sometimes the system has too many unknowns to easily construct and interpret a root locus plot (e.g. with two PID controllers there are the variables Kc1, Kc2, τi1, τi2, τd1, and τd2).

Chapter 4

PROGRAMMABLE LOGIC CONTROLLER

A **programmable logic controller, PLC**, or **programmable controller** is a digital computer used for automation of typically industrial electromechanical processes, such as control of machinery on factory assembly lines, amusement rides, or light fixtures. PLCs are used in many machines, in many industries.

PLCs are designed for multiple arrangements of digital and analog inputs and outputs, extended temperature ranges, immunity to electrical noise, and resistance to vibration and impact. Programs to control machine operation are typically stored in battery-backed-up or non-volatile memory. A PLC is an example of a "hard" real-time system since output results must be produced in response to input conditions within a limited time, otherwise unintended operation will result.

Before the PLC, control, sequencing, and safety interlock logic for manufacturing automobiles was mainly composed of relays, cam timers, drum sequencers, and dedicated closed-loop controllers. Since these could number in the hundreds or even thousands, the process for updating such facilities for the yearly model change-over was very time consuming and expensive, as electricians needed to individually rewire the relays to change their operational characteristics.

Digital computers, being general-purpose programmable devices, were soon applied to control of industrial processes. Early computers required specialist programmers, and stringent operating environmental control for temperature, cleanliness, and power quality. Using a general-purpose computer for process control required protecting the computer from the plant floor conditions. An industrial control computer would have several attributes: it would tolerate the shop-floor environment, it would support discrete (bit-form) input and output in an easily extensible manner, it would not require years of training to use, and it would permit its operation to be monitored.

The response time of any computer system must be fast enough to be useful for control; the required speed varying according to the nature of the process.

Since many industrial processes have timescales easily addressed by millisecond response times, modern (fast, small, reliable) electronics greatly facilitate building reliable controllers, especially because performance can be traded off for reliability.

In 1968 GM Hydra-Matic(the automatic transmission division of General Motors) issued a request for proposals for an electronic replacement for hard-wired relay systems based on a white paper written by engineer Edward R. Clark. The winning proposal came from Bedford Associates of Bedford, Massachusetts. The first PLC, designated the 084 because it was Bedford Associates' eighty-fourth project, was the result.

Bedford Associates started a new company dedicated to developing, manufacturing, selling, and servicing this new product: Modicon, which stood for **MO**dular**DI**gital**CON**troller. One of the people who worked on that project was Dick Morley, who is considered to be the "father" of the PLC. The Modicon brand was sold in 1977 to Gould Electronics, later acquired by German Company AEG, and then by French Schneider Electric, the current owner.

One of the very first 084 models built is now on display at Modicon's headquarters in North Andover, Massachusetts. It was presented to Modicon by GM, when the unit was retired after nearly twenty years of uninterrupted service. Modicon used the 84 moniker at the end of its product range until the 984 made its appearance.

The automotive industry is still one of the largest users of PLCs.

DEVELOPMENT

Early PLCs were designed to replace relay logic systems. These PLCs were programmed in "ladder logic", which strongly resembles a schematic diagram of relay logic. This program notation was chosen to reduce training demands for the existing technicians. Other early PLCs used a form of instruction list programming, based on a stack-based logic solver.

Modern PLCs can be programmed in a variety of ways, from the relay-derived ladder logic to programming languages such as specially adapted dialects of BASIC and C. Another method is state logic, a very high-level programming language designed to program PLCs based on state transition diagrams.

Many early PLCs did not have accompanying programming terminals that were capable of graphical representation of the logic, and so the logic was instead represented as a series of logic expressions in some version of Boolean format, similar to Boolean algebra. As programming terminals evolved, it became more common for ladder logic to be used, for the aforementioned reasons and because it was a familiar format used for electromechanical control panels.

Newer formats such as state logic and Function Block (which is similar to the way logic is depicted when using digital integrated logic circuits) exist, but they are still not as popular as ladder logic. A primary reason for this is that PLCs solve the logic in a predictable and repeating sequence, and ladder logic allows the programmer (the person writing the logic) to see any issues with the timing of the logic sequence more easily than would be possible in other formats.

Programming

Early PLCs, up to the mid-1990s, were programmed using proprietary programming panels or special-purpose programming terminals, which often had dedicated function keys representing the various logical elements of PLC programs. Some proprietary programming terminals displayed the elements of PLC programs as graphic symbols, but plain ASCII character representations of contacts, coils, and wires were common. Programs were stored on cassette tape cartridges. Facilities for printing and documentation were minimal due to lack of memory capacity. The oldest PLCs used non-volatile magnetic core memory.

More recently, PLCs are programmed using application software on personal computers, which now represent the logic in graphic form instead of character symbols. The computer is connected to the PLC through Ethernet, RS-232, RS-485, or RS-422 cabling. The programming software allows entry and editing of the ladder-style logic.

Generally the software provides functions for debugging and troubleshooting the PLC software, for example, by highlighting portions of the logic to show current status during operation or via simulation. The software will upload and download the PLC program, for backup and restoration purposes. In some models of programmable controller, the program is transferred from a personal computer to the PLC through a programming board which writes the program into a removable chip such as an EPROM

Functionality

The functionality of the PLC has evolved over the years to include sequential relay control, motion control, process control, distributed control systems, and networking. The data handling, storage, processing power, and communication capabilities of some modern PLCs are approximately equivalent to desktop computers. PLC-like programming combined with remote I/O hardware, allow a general-purpose desktop computer to overlap some PLCs in certain applications.

Desktop computer controllers have not been generally accepted in heavy industry because the desktop computers run on less stable operating systems than do PLCs, and because the desktop computer hardware is typically not designed to the same levels of tolerance to temperature, humidity, vibration, and longevity as the processors used in PLCs.

Operating systems such as Windows do not lend themselves to deterministic logic execution, with the result that the controller may not always respond to changes of input status with the consistency in timing expected from PLCs. Desktop logic applications find use in less critical situations, such as laboratory automation and use in small facilities where the application is less demanding and critical, because they are generally much less expensive than PLCs.

Programmable Logic Relay (PLR)

In more recent years, small products called PLRs (programmable logic relays), and also by similar names, have become more common and accepted. These are much like PLCs, and are used in light industry where only a few points of I/O(*i.e.* a few signals coming in from the real world and a few going out) are needed, and low cost is desired. These small devices are typically made in a common physical size and shape by several manufacturers, and branded by the makers of larger PLCs to fill out their low end product range.

Popular names include PICO Controller, NANO PLC, and other names implying very small controllers. Most of these have 8 to 12 discrete inputs, 4 to 8 discrete outputs, and up to 2 analog inputs. Size is usually about 4" wide, 3" high, and 3" deep. Most such devices include a tiny postage-stamp-sized LCD screen for viewing simplified ladder logic (only a very small portion of the program being visible at a given time) and status of I/O points, and typically these screens are accompanied by a 4-way rocker push-button plus four more separate push-buttons, similar to the key buttons on a VCR remote control, and used to navigate and edit the logic.

Most have a small plug for connecting via RS-232 or RS-485 to a personal computer so that programmers can use simple Windows applications for programming instead of being forced to use the tiny LCD and push-button set for this purpose. Unlike regular PLCs that are usually modular and greatly expandable, the PLRs are usually not modular or expandable, but their price can be two orders of magnitude less than a PLC, and they still offer robust design and deterministic execution of the logics.

PLC Topics

Features

The main difference from other computers is that PLCs are armored for severe conditions (such as dust, moisture, heat, cold), and have the facility for extensive input/output(I/O) arrangements. These connect the PLC to sensors and actuators. PLCs read limit switches, analog process variables (such as temperature and pressure), and the positions of complex positioning systems. Some use machine vision.

On the actuator side, PLCs operate electric motors, pneumatic or hydraulic cylinders, magnetic relays, solenoids, or analog outputs. The input/output arrangements may be built into a simple PLC, or the PLC may have external I/O modules attached to a computer network that plugs into the PLC.

Fig. : Control panel with PLC (grey elements in the center). The unit consists of separate elements, from left to right; power supply, controller, relay units for in- and output.

Scan Time

A PLC program is generally executed repeatedly as long as the controlled system is running. The status of physical input points is copied to an area of memory accessible to the processor, sometimes called the "I/O Image Table". The program is then run from its first instruction rung down to the last rung. It takes some time for the processor of the PLC to evaluate all the rungs and update the I/O image table with the status of outputs.

This scan time may be a few milliseconds for a small program or on a fast processor, but older PLCs running very large programs could take much longer (say, up to 100ms) to execute the program. If the scan time were too long, the response of the PLC to process conditions would be too slow to be useful.

As PLCs became more advanced, methods were developed to change the sequence of ladder execution, and subroutines were implemented. This simplified programming could be used to save scan time for high-speed processes; for example, parts of the program used only for setting up the machine could be segregated from those parts required to operate at higher speed.

Special-purpose I/O modules may be used where the scan time of the PLC is too long to allow predictable performance. Precision timing modules, or counter modules for use with shaft encoders, are used where the scan time would be too long to reliably count pulses or detect the sense of rotation of an encoder. The relatively slow PLC can still interpret the counted values to control a machine, but the accumulation of pulses is done by a dedicated module that is unaffected by the speed of the program execution.

System Scale

A small PLC will have a fixed number of connections built in for inputs and outputs. Typically, expansions are available if the base model has insufficient I/O.

Modular PLCs have a chassis (also called a rack) into which are placed modules with different functions. The processor and selection of I/O modules are customized for the particular application. Several racks can be administered by a single processor, and may have thousands of inputs and outputs. Either a special high speed serial I/O link or comparable communication method is used so that racks can be distributed away from the processor, reducing the wiring costs for large plants. Options are also available to mount I/O points directly to the machine and utilize quick disconnecting cables to sensors and valves, saving time for wiring and replacing components.

User Interface

PLCs may need to interact with people for the purpose of configuration, alarm reporting, or everyday control. A human-machine interface(HMI) is employed for this purpose. HMIs are also referred to as man-machine interfaces (MMIs) and graphical user interfaces (GUIs). A simple system may use buttons and lights to interact with the user. Text displays are available as well as graphical touch screens. More complex systems use programming and monitoring software installed on a computer, with the PLC connected via a communication interface.

Communications

PLCs have built-in communications ports, usually 9-pin RS-232, RS-422, rs-485, Ethernet. Various protocols are usually included. Many of these protocols are vendor specific.

Most modern PLCs can communicate over a network to some other system, such as a computer running a SCADA(Supervisory Control And Data Acquisition) system or web browser.

PLCs used in larger I/O systems may have peer-to-peer(P2P) communication between processors. This allows separate parts of a complex process to have individual control while allowing the subsystems to co-ordinate over the communication link. These communication links are also often used for HMI devices such as keypads or PC-type workstations.

Formerly, some manufacturers offered dedicated communication modules as an add-on function where the processor had no network connection built-in.

Programming

PLC programs are typically written in a special application on a personal computer, then downloaded by a direct-connection cable or over a network to the PLC. The program is stored in the PLC either in battery-backed-up RAM or some other non-volatile flash memory. Often, a single PLC can be programmed to replace thousands of relays.

Under the IEC 61131-3 standard, PLCs can be programmed using standards-based programming languages. A graphical programming notation called Sequential Function Charts is available on certain programmable controllers. Initially most PLCs utilized Ladder Logic Diagram Programming, a model which emulated electromechanical control panel devices (such as the contact and coils of relays) which PLCs replaced. This model remains common today.

IEC 61131-3 currently defines five programming languages for programmable control systems: function block diagram(FBD), ladder diagram(LD), structured text(ST; similar to the Pascal programming language), instruction list(IL; similar to assembly language), and sequential function chart(SFC). These techniques emphasize logical organization of operations.

While the fundamental concepts of PLC programming are common to all manufacturers, differences in I/O addressing, memory organization, and instruction sets mean that PLC programs are never perfectly interchangeable between different makers. Even within the same product line of a single manufacturer, different models may not be directly compatible.

Security

Prior to the discovery of the Stuxnetcomputer worm in June 2010, security of PLCs received little attention. PLCs generally contain a real-time operating system such as OS-9 or VxWorks, and exploits for these systems exist much as they do for desktop computer operating systems such as Microsoft Windows. PLCs can also be attacked by gaining control of a computer they communicate with.

Simulation

In order to properly understand the operation of a PLC, it is necessary to spend considerable time programming, testing, and debugging PLC programs. PLC systems are inherently expensive, and down-time is often very costly. In addition, if a PLC is programmed incorrectly it can result in lost productivity and dangerous conditions. PLC simulation software such as PLCLogix can save time in the design of automated control applications and can also increase the level of safety associated with equipment since various "what if" scenarios can be tried and tested before the system is activated.

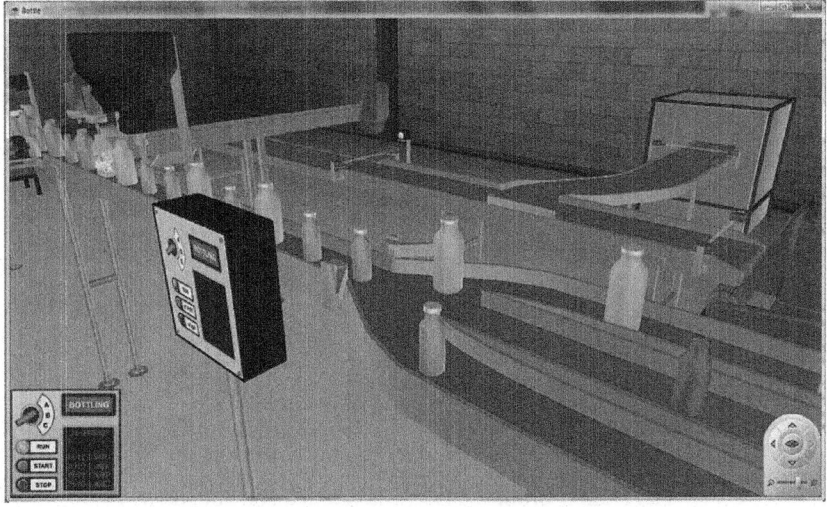

Fig. : PLCLogix PLC Simulation Software.

Redundancy

Some special processes need to work permanently with minimum unwanted down time. Therefore, it is necessary to design a system which is fault-tolerant and capable of handling the process with faulty modules. In such cases to increase the system availability in the event of hardware component failure, redundant CPU or I/O modules with the same functionality can be added to hardware configuration for preventing total or partial process shutdown due to hardware failure.

PLC Compared with Other Control Systems

PLCs are well adapted to a range of automation tasks. These are typically industrial processes in manufacturing where the cost of developing and maintaining the automation system is high relative to the total cost of the automation, and where changes to the system would be expected during its operational life.

PLCs contain input and output devices compatible with industrial pilot devices and controls; little electrical design is required, and the design problem centers on expressing the desired sequence of operations.

PLC applications are typically highly customized systems, so the cost of a packaged PLC is low compared to the cost of a specific custom-built controller design. On the other hand, in the case of mass-produced goods, customized control systems are economical. This is due to the lower cost of the components, which can be optimally chosen instead of a "generic" solution, and where the non-recurring engineering charges are spread over thousands or millions of units.

Fig. : Allen-Bradley PLC installed in a control panel.

For high volume or very simple fixed automation tasks, different techniques are used. For example, a consumer dishwasher would be controlled by an electro-mechanical cam timer costing only a few dollars in production quantities.

A microcontroller-based design would be appropriate where hundreds or thousands of units will be produced and so the development cost (design of power supplies, input/output hardware, and necessary testing and certification) can be spread over many sales, and where the end-user would not need to alter the control.

Automotive applications are an example; millions of units are built each year, and very few end-users alter the programming of these controllers. However, some specialty vehicles such as transit buses economically use PLCs instead of custom-designed controls, because the volumes are low and the development cost would be uneconomical.

Very complex process control, such as used in the chemical industry, may require algorithms and performance beyond the capability of even high-perfor-mance PLCs. Very high-speed or precision controls may also require customized solutions; for example, aircraft flight controls. Single-board computers using semi-customized or fully proprietary hardware may be chosen for very demanding control applications where the high development and maintenance cost can be supported. "Soft PLCs" running on desktop-type computers can interface with industrial I/O hardware while executing programs within a version of commercial operating systems adapted for process control needs.

Programmable controllers are widely used in motion control, positioning control, and torque control. Some manufacturers produce motion control units to be integrated with PLC so that G-code(involving a CNC machine) can be used to instruct machine movements.

PLCs may include logic for single-variable feedback analog control loop, a proportional, integral, derivative (PID) controller. A PID loop could be used to control the temperature of a manufacturing process, for example. Historically PLCs were usually configured with only a few analog control loops; where processes required hundreds or thousands of loops, a distributed control system(DCS) would instead be used. As PLCs have become more powerful, the boundary between DCS and PLC applications has become less distinct.

PLCs have similar functionality as remote terminal units(RTU). An RTU, however, usually does not support control algorithms or control loops. As hardware rapidly becomes more powerful and cheaper, RTUs, PLCs, and DCSs are increasingly beginning to overlap in responsibilities, and many vendors sell RTUs with PLC-like features, and vice versa. The industry has standardized on the IEC 61131-3 functional block language for creating programs to run on RTUs and PLCs, although nearly all vendors also offer proprietary alternatives and associated development environments.

In recent years "safety" PLCs have started to become popular, either as standalone models or as functionality and safety-rated hardware added to existing controller architectures (Allen Bradley Guardlogix, Siemens F-series *etc.*. These differ from conventional PLC types as being suitable for use in safety-critical applications for which PLCs have traditionally been supplemented with hard-wired safety relays.

For example, a safety PLC might be used to control access to a robot cell with trapped-key access, or perhaps to manage the shutdown response to an emergency stop on a conveyor production line. Such PLCs typically have a restricted regular instruction set augmented with safety-specific instructions designed to interface with emergency stops, light screens, and so forth. The flexibility that such systems offer has resulted in rapid growth of demand for these controllers.

Discrete and Analog Signals

Discrete signals behave as binary switches, yielding simply an On or Off signal (1 or 0, True or False, respectively). Push buttons, limit switches, and photoelectric sensors are examples of devices providing a discrete signal. Discrete signals are sent using either voltage or current, where a specific range is designated as *On* and another as *Off*. For example, a PLC might use 24 V DC I/O, with values above 22 V DC representing *On*, values below 2VDC representing *Off*, and intermediate values undefined. Initially, PLCs had only discrete I/O.

Analog signals are like volume controls, with a range of values between zero and full-scale. These are typically interpreted as integer values (counts) by the PLC, with various ranges of accuracy depending on the device and the number of bits available to store the data. As PLCs typically use 16-bit signed binary processors, the integer values are limited between -32,768 and +32,767.

Pressure, temperature, flow, and weight are often represented by analog signals. Analog signals can use voltage or current with a magnitude proportional to the value of the process signal. For example, an analog 0 to 10 V or 4-20 mA input would be converted into an integer value of 0 to 32767.

Current inputs are less sensitive to electrical noise (*e.g.* from welders or electric motor starts) than voltage inputs.

Example

As an example, say a facility needs to store water in a tank. The water is drawn from the tank by another system, as needed, and our example system must manage the water level in the tank by controlling the valve that refills the tank. Shown is a "ladder diagram" which shows the control system. A ladder diagram is a method of drawing control circuits which pre-dates PLCs. The ladder diagram resembles the schematic diagram of a system built with electromechanical relays. Shown are:

- Two inputs (from the low and high level switches) represented by contacts of the float switches
- An output to the fill valve, labelled as the fill valve which it controls
- An "internal" contact, representing the output signal to the fill valve which is created in the program.
- A logical control scheme created by the interconnection of these items in software

In ladder diagram, the contact symbols represent the state of bits in processor memory, which corresponds to the state of physical inputs to the system. If a discrete input is energized, the memory bit is a 1, and a "normally open" contact controlled by that bit will pass a logic "true" signal on to the next element of the ladder. Therefore, the contacts in the PLC program that "read" or look at the physical switch contacts in this case must be "opposite" or open in order to return a TRUE for the closed physical switches. Internal status bits, corresponding to the state of discrete outputs, are also available to the program.

In the example, the physical state of the float switch contacts must be considered when choosing "normally open" or "normally closed" symbols in the ladder diagram. The PLC has two discrete inputs from float switches(Low Level and High Level). Both float switches (normally closed) open their contacts when the water level in the tank is above the physical location of the switch.

When the water level is below both switches, the float switch physical contacts are both closed, and a true (logic 1) value is passed to the Fill Valve output. Water begins to fill the tank. The internal "Fill Valve" contact latches the circuit so that even when the "Low Level" contact opens (as the water passes the lower switch), the fill valve remains on. Since the High Level is also normally closed, water continues to flow as the water level remains between the two switch levels.

Once the water level rises enough so that the "High Level" switch is off (opened), the PLC will shut the inlet to stop the water from overflowing; this is an example of seal-in (latching) logic. The output is sealed in until a high level condition breaks the circuit. After that the fill valve remains off until the level drops so low that the Low Level switch is activated, and the process repeats again.

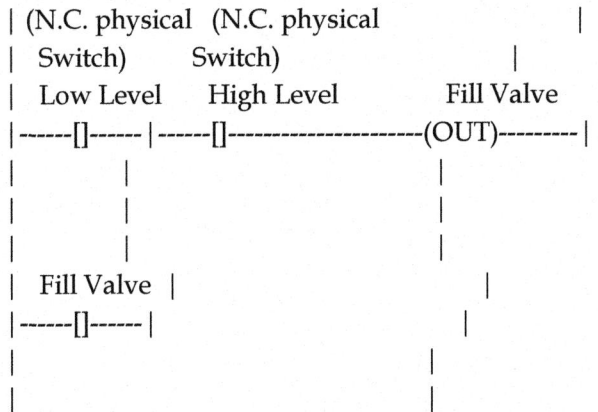

A complete program may contain thousands of rungs, evaluated in sequence. Typically the PLC processor will alternately scan all its inputs and update outputs, then evaluate the ladder logic; input changes during a program scan will not be effective until the next I/O update. A complete program scan may take only a few milliseconds, much faster than changes in the controlled process.

Programmable controllers vary in their capabilities for a "rung" of a ladder diagram. Some only allow a single output bit. There are typically limits to the number of series contacts in line, and the number of branches that can be used. Each element of the rung is evaluated sequentially. If elements change their state during evaluation of a rung, hard-to-diagnose faults can be generated, although sometimes (as above) the technique is useful. Some implementations forced evaluation from left-to-right as displayed and did not allow reverse flow of a logic signal (in multi-branched rungs) to affect the output.

LOGICAL MODELING

Boolean Models

A Boolean is a variable that can only attain two values: True or False. In most applications, it is convenient to represent a True by the number 1, and a False by the number 0. A **Boolean model**, or **Boolean network**, is a collection of Boolean variables that are related by logical switching rules, or **Boolean functions**, that follow an If-Then format. This type of Boolean model is known as an autonomous model and will be the primary type of model discussed in this article.

In chemical engineering, Boolean models can be used to model simple control systems. Boolean functions can be used to model switches on pumps and valves that react to readings from sensors that help keep that system operating smoothly and safely.

A simple application for level control of a CSTR is included in worked-out example 1. In this example, Boolean function is used to close the inlet stream and open the outlet stream when the level is higher than a specified point.

Boolean Functions

Boolean functions are logical operators that relate two or more Boolean variables within a system and return a true or false. A Boolean expression is a group of Boolean functions, which will be described individually below. When computing the value of a Boolean expression, Parentheses are used to indicate priority (working from inside out as in algebra). After that, LOGICAL INVERSION will always be first and LOGICAL EQUIVALENCE will be last, but the order of operation for the AND, OR, and EXCLUSIVE OR functions are specified with parenthesis.

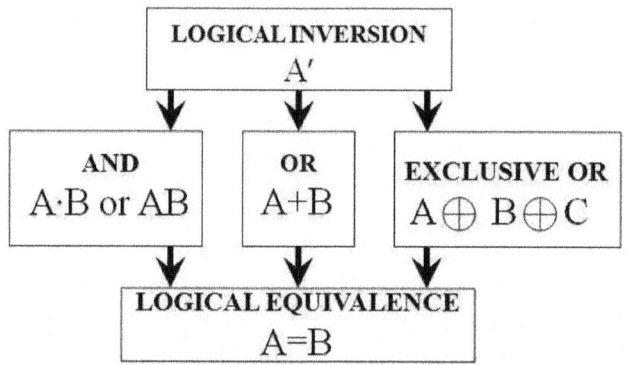

Descriptions and examples of these functions are given below. A quick reference of each of the functions can be found after the examples.

Logical Inversion

LOGICAL INVERSION is a function that returns the opposite value of a variable. The function is denoted as a prime on the variable (*e.g.* A' or B')For example, if we say that A is true (A=1), then the function A' will return a false (A'=0). Similarly, if we say that A is false (A=0) then the function A' will return true (A'=1).

Example:

A=1, B=A' then B=0

And

The AND function relates two or more Boolean variables and returns a true if-and-only-if both variables are true. A dot is used to denote the AND function, or it is simply omitted. For example "A and B" can be written as "A•B" or as "AB." In this example, the AND function will only return a true if-and-only-if both Boolean variables A and B have a value of 1.

Examples:

Variables	Results
A=1, B=1	AB = 1
A=1, B=0	AB = 0
A=1, B=1, C=1	ABC = 1
A=1, B=0, C=1	ABC = 0

Or

The OR function relates two or more Boolean variables and returns a true if any referenced variables are true. A plus is used to denote the OR function. For example "A or B" can be written as "A+B." In this example, the OR function will return true if either Boolean variable, A or B, has a value of 1.

Examples:

Variables	Results
A=1, B=1	A+B = 1
A=1, B=0	A+B = 1
A=0, B=0	A+B = 0
A=0, B=0, C=1	A+B+C = 1
A=0, B=0, C=0	A+B+C = 0

Exclusive Or

The EXCLUSIVE OR function relates two or more Boolean variables and returns **true** only when **one** of the variables is true and **all** other variables are false. It returns **false** when **more than one** of the variables are true, or **all** the variables are false. A circumscribed plus is used to denote the EXCLUSIVE OR function. For example "A EXCLUSIVE OR B" can be written as "A $\oplus$ B."

Examples:

Variables	Results
A=1, B=1	A $\oplus$ B = 0
A=1, B=0	A $\oplus$ B = 1

A=0, B=1	$A \oplus B = 1$
A=0, B=0	$A \oplus B = 0$
A=0, B=0, C=0	$A \oplus B \oplus C = 0$
A=1, B=0, C=0	$A \oplus B \oplus C = 1$
A=1, B=0, C=1	$A \oplus B \oplus C = 0$
A=1, B=1, C=1	$A \oplus B \oplus C = 0$

Logical Equivalence

The LOGICAL EQUIVALENCE function equates two Boolean variables or expressions. The LOGICAL EQUIVALENCE function, denoted as =, assigns a Boolean variable a true or false depending on the value of the variable or expression that it is being equated with. For example, A LOGICAL EQUIVALENCE B can be written as A = B. In this example, the value of A will be assigned the value of B.

Boolean Networks

As stated in the introduction, a Boolean network is a system of boolean equations. In chemical engineering, Boolean networks are likely to be dependant on external inputs as a means of controlling a physical system. However, the following sections pertain mostly to synchronous autonomous systems.

An autonomous system is one that is completely independent of external inputs. Every Boolean variable is dependent on the state of other Boolean variables in the system and no variable is controlled by an external input. A synchronous system is one that logical switching (the changing of Boolean variables) occurs simultaneously for all variables based on the values prior to the incidence of change.

Here is an example of an autonomous boolean network:

Boolean Functions
$A = B+C'$
$B = AC$
$C = A'$

Truth Tables

A truth table is a tabulation of all the possible states of a Boolean Model at different time frames. A simple truth table shows the potential initial states at time, T_i, and the corresponding subsequent states at time T_{i+1}, of a Boolean network. Truth tables can provide one with a clearer picture of how the rules apply and how they affect each situation. Hence, they help to ensure that each output only has one control statement so that the Boolean rules do not conflict with each other.

Constructing Truth Tables

1. Draw up a table with the appropriate number of columns for each variable; one for each input and output.

2. The left side of the column should contain all possible permutations of the input variables at time T_i. One method to accomplish this might be to list all possible combinations in ascending binary order.

3. The right side of the column should contain the corresponding outcome of the output variables at the subsequent time T_{i+1}. A generic example of this with 2 variables can be seen below:

Inputs		Outputs	
T_i		T_{i+1}	
variable1	variable2	variable1	variable2

A quick way to check that you have all of the possible permutations is that there should be 2^x possible permutations for X input variables.

Example of a Truth Table

The sample system we will be using is based on hydrogen fuel cell technology. The equation for the operation of hydrogen fuel cells is

$$H_2 + O_2 \rightarrow H_2O.$$

One aspect of Proton Exchange Membrane (PEM) fuel cells (a type of fuel cell) is that the performance of the fuel cell is highly dependent on the relative humidity of the system (if humidity rises too high, the fuel cell will flood and H_2 and O_2 will be unable to reach the cell. If humidity falls too low, the fuel cell will dry up and the performance will drop. The task is to create a Boolean model for this simplified water management system.

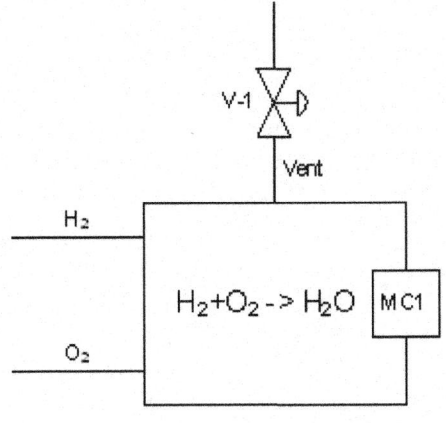

PEM Fuel Cell System

The system produces steam within the system, and there is a vent to release steam if the system becomes too saturated. In our system, we will assume that the inputs are stoichimetric and react completely. Also we will assume that pressure buildup from steam is negligible compared to the change in relative humidity. The only variable in question is the %relative humidity in the system.

* note: this is not how water management actually works in a fuel cell system, but it is a simple example.

A will represent the moisture controller response (0 indicates relative humidity or %RH <80%, 1 indicates %RH >80%) B will represent the valve status (0 is closed, 1 is open)

The corresponding Boolean functions for this model are given below (normally you would have to design these yourself to meet the criteria you desire):

A = A

B = A

For this example with 2 input variables, there are $2^2 = 4$ possible permutations and $2^2 = 4$ rows. The resultant permutations for the outputs are: For A where Y=1, the number of 0s and 1s are $2^{(Y-1)}=2^{(1-1)}=1$. For B where Y=2, the number of 0s and 1s are $2^{(Y-1)}=2^{(2-1)}=2$.

The resultant truth table is below:

Inputs		Outputs	
T_i		T_{i+1}	
A	B	A	B
0	0	0	0
1	0	1	1
0	1	0	0
1	1	1	1

State Transition Diagrams

A state transition diagram is a graphical way of viewing truth tables. This is accomplished by looking at each individual initial state and its resultant state. The transition from one state to another is represented by an arrow. Then they are pieced together like a jigsaw puzzle until they fit in place. When one state leads to itself it simply points to itself. The following example is based on the truth table in the previous section.

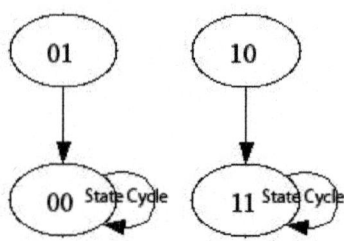

In this example, there are two state cycles. A state cycle is a combination of states around which the system continually enters and reenters. For a finite number of states, there will always exist at least one state cycle. A state cycle is also a pathway or a flowchart that shows the "decision making process" of a Boolean network. This feature is a direct result from two attributes of Boolean networks:

1. Finite number of states

2. Deterministic (there is a certain set of rules that determines the next state that will be entered)

In the example presented in the previous section, there were two state cycles. One advantage of state cycles is it easily allows you to see where your model will end up cycling and if there are any states that are not accounted for properly by your model. In the previous diagram, if the moisture controller indicated the humidity was below the set value, it would close the valve or hold the valve closed. If the moisture controller indicated that the humidity was above the set value, it would either open the valve or hold it open.

Consider this alternate system.

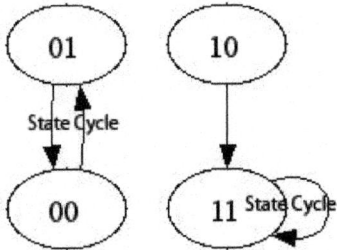

In this example, the state cycle says that if the meter says that the humidity is below the set point it would cycle the vent valve open and closed. This would hurt the system and is not a desired outcome of the model.

For safety and functionality issues, a process control engineer would want to consider all possiblities in the design of any Boolean network modeling a real system.

Limitations of Boolean Networks

Advantages of Boolean Networks

- Unlike ordinary differential equations and most other models, Boolean networks do not require an input of parameters.
- Boolean models are quick and easy to compute using computers.
- Boolean networks can be used to model a wide variety of activities and events.
- Boolean networks can be used to approximate ordinary differential equations when there are an infinite number of states.

Disadvantages of Boolean Networks

- Boolean networks are restrained to computing very simple math. They cannot be used for calculus and to calculate large quantities.
- Boolean models have relatively low resolution compared to other models.
- It is very time consuming and complicated to build Boolean networks by hand.

Example 1

A hypothetical CSTR needs to have its liquid level maintained below a safety mark by means of a sensor, L1, on the corresponding mark and a control valve placed on the inlet and outlet streams – V1 and V2 respectively. A typical application of the afore-mentioned system could involve heterogeneously catalyzed liquid reaction(s) with liquid product(s).

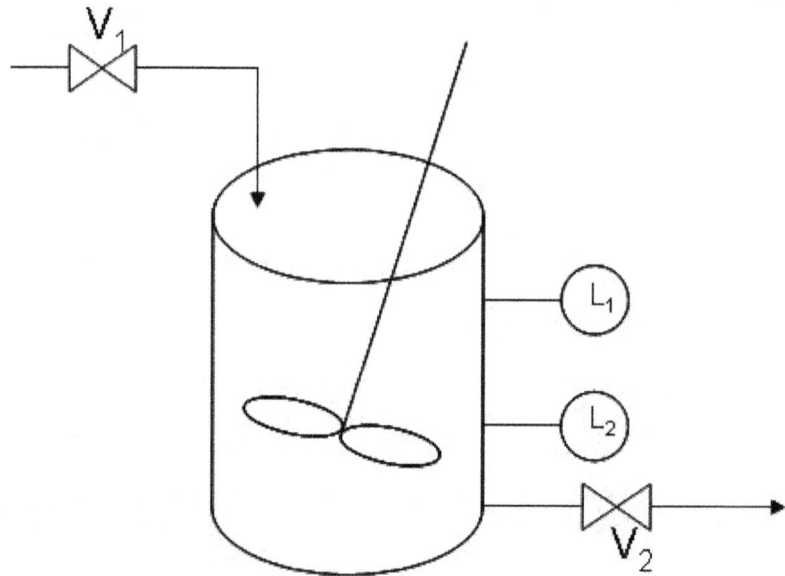

Solution to Example

Conventions

Water level sensor

L1	0	1
water level	desirable	too high

Valve

V	0	1
position	closed	open

Initial State

Assume that the CSTR is empty and being filled up. CSTR, being empty, sets the value of L1 to zero. Filling up the CSTR could be done by opening valve 1 - V1 assuming a value of one - and closing valve 2 - V2 assuming a value of zero.

In coordinate form, the initial state is as such: $(L1, V1, V2) = (0, 1, 0)$

Problem Interpretation

Let h be the water level and WL1 be the safety mark defined in the CSTR. The system could assume one of the following states at any one time:

1) h< WL1 : desirable water level

Maximizing production of the chemical prompts the system to remain in its current state - that is, its initial state.

$(L1, V1, V2)$final $= (0, 1, 0)$**final state**

2) h> WL1 : water level too high

Prevention of flooding requires that the tank be emptied. As such, valve 1(V1) should be closed to stop the input while valve 2(V2) should be open to empty the extra water above the safety water mark.

$(L1, V1, V2)' = (1, 1, 0)$**trigger to valve**

$(L1, V1, V2)$final $= (1, 0, 1)$**final state**

State Cycle

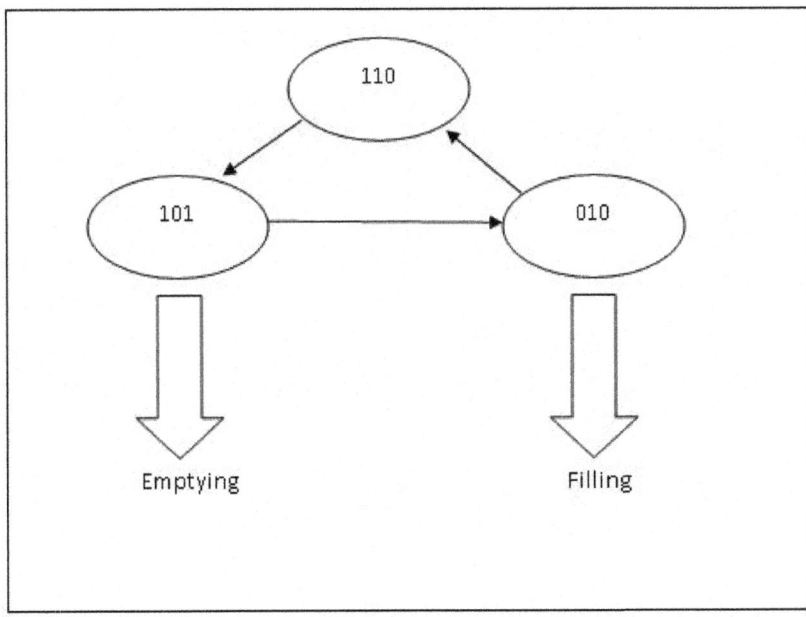

Physical Significance

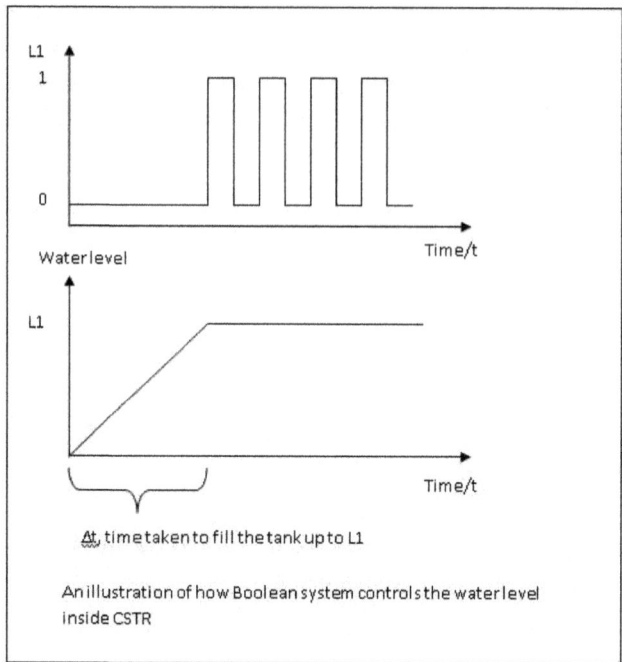

An illustration of how Boolean system controls the water level inside CSTR

Quick Reference

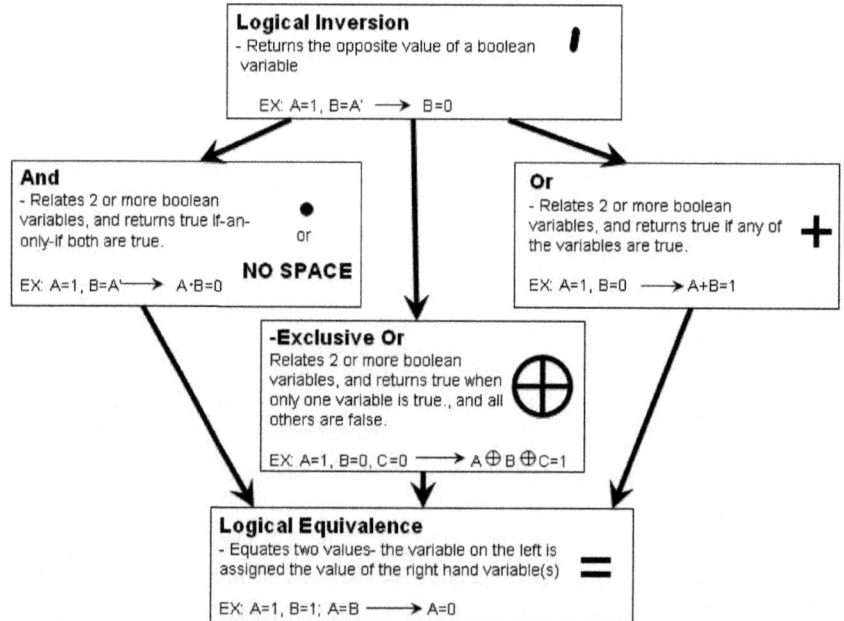

LOGICAL PROGRAMS

A logical control program is a set of conditional statements describing the response of a controller to different inputs. A controller is a computer used to automate industrial processes. Process engineers use control logic to tell the controller in a process how to react to all inputs from sensors with an appropriate response to maintain normal functioning of the process.

Control logic (sometimes called process logic) is based on simple logic principles governed by statements such as IF X, THEN Y, ELSE Z yet can be used to describe a wide range of complex relationships in a process. Although different controllers and processes use different programming languages, the concepts of control logic apply and the conditions expressed in a logical control program can be adapted to any language.

The concepts behind logical control programs are not only found in chemical processes; in fact, control logic is used in everyday life. For example, a person may regulate his/her own body temperature and comfort level using the following conditional logic statements: IF the temperature is slightly too warm, THEN turn on a fan; IF the temperature is way too warm, THEN turn on the air conditioning; IF the temperature is slightly too cold, THEN put on a sweatshirt; IF the temperature is way too cold, THEN turn on the fireplace. The person takes an input from the environment (temperature) and if it meets a certain prescribed condition, she executes an action to keep herself comfortable. Similarly, chemical processes evaluate input values from the process against set values to determine the necessary actions to keep the process running smoothly and safely (aforementioned example illustrated below).

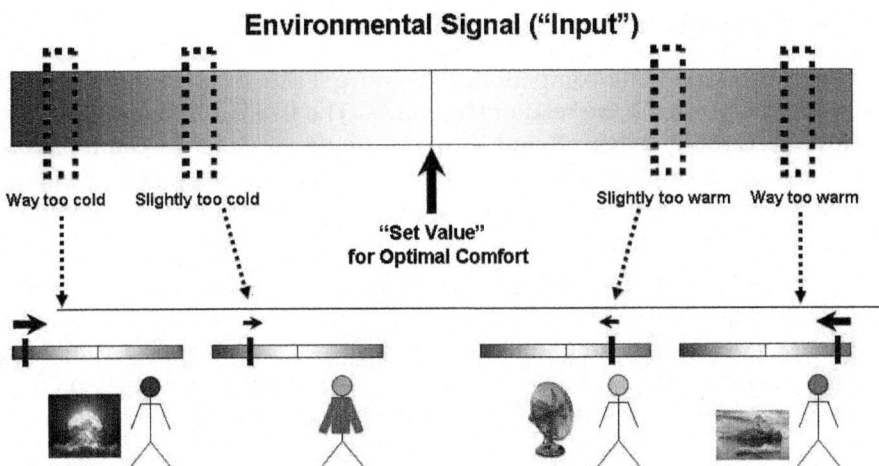

The following sections elaborate on the construction of conditional logic statements and give examples of developing logical control programs for chemical processes.

Logic Controls

Logical controls (IF, THEN, ELSE, and WHILE) compare a value from a sensor to a set standard for the value to evaluate the variable as True/False in order to dictate an appropriate response for the physical system. The control program for a chemical process contains many statements describing the responses of valves, pumps, and other equipment to sensors such as flow and temperature sensors. The responses described by the system can be discrete, such as an on/off switch, or can be continuous, such as opening a valve between 0 and 100%.

The goal of a control program is to maintain the values monitored by the sensors at an acceptable level for process operation considering factors like product quality, safety, and physical limitations of the equipment. In addition to describing the normal activity of the process, a control program also describes how the process will initialize at the start of each day and how the controller will respond to an emergency outside of the normal operating conditions of the system.

Unlike a linear computer program, logic programs are continuously monitoring and responding without a specific order. Before constructing logical control programs, it is important to understand the conditional statements, such as IF-THEN and WHIIE statements, that govern process logic.

The following logic controls found below are written in pseudocode. Pseudocode is a compact and informal way of writing computer program algorithms. It is intended for human reading instead of machine reading, and does not require stringent syntax for people to understand. Pseudocode is typically used for planning computer program development and to outline a program before actual coding occurs.

IF-THEN Statements

IF-THEN statements compare a value from a sensor to a set value and describe what should happen if the relationship holds. The IF-THEN statement takes the form IF X, THEN Y where X and Y can be single variables or combinations of variables. For example, consider the following statements 1 and 2. In statement 1, both X and Y are single variables whereas in statement 2, X is a combination of two variables.

The ability in conditional logic to combine different conditions makes it more flexible than incidence graphs, which can only describe monotonic relationships between two variables. A monotonic relationship is one where if X is increasing, Y is always decreasing or if X is increasing, Y is always increasing. For complex processes, it is important to be able to express non-monotonic relationships.

1. IF T>200 C, THEN open V1
2. IF T>200 C and P>200 psi, THEN open V1.

Where T is Temperature, P is pressure, and V represents a valve.

In statement 1, if the temperature happens to be above 200C, valve 1 will be opened. In statement 2, if the temperature is above 200 C and the pressure is above 200 psi, then the valve will be opened.

Otherwise, no action will be taken on valve 1

If the conditions in the IF statement are met, the THEN statement is executed, and depending on the command, the physical system is acted upon. Otherwise, no action is taken in response to the sensor input. In order to describe an alternate action if the IF condition does not hold true, ELSE statements are necessary.

ELSE Statements

The simple form of an IF-THEN-ELSE statement is IF X, THEN Y, ELSE Z where again X, Y, and Z can be single variables or combinations of variables (as explained in the IF-THEN section above). The variable(s) in the ELSE statement are executed if the conditions in the IF statement are not true. This statement works similar to the IF-THEN statements, in that the statements are processed in order. The ELSE statement is referred to last, and is a condition that is often specified to keep the program running. An example is the following:

IF P>200 psi, THEN close V1

ELSE open V4

In this statement, if the pressure happens to be 200psi or less, the THEN statement will be skipped and the ELSE statement will be executed, opening valve 4.

Sometimes, if X, Y or Z represent many variables and several AND or OR statements are used, a WHILE statement may be employed.

CASE Statements

CASE statement is an alternative syntax that can be cleaner than many IF.. THEN and ELSE statements. The example shown in the table below shows its importance.

CASE	IF..THEN and ELSE

T>Tset+2:
     v2=v2+0.1
T>Tset+1:
     v2=v2+0.05
T<Tset-1:
     v2=v2-0.05
T<Tset-2:
     v2=v2-0.1
	IF T>Tset+1:
	     IF T>Tset+2:
	     v2=v2+0.1
	     ELSE:
	     v2=v2+0.05
	ELSE IF T<Tset-1:
	     IF T<Tset-2:
	     v2=v2-0.1
	     ELSE:
	     v2=v2-0.05

Thus CASE statements make the code easier to read for debugging.

WHILE Statements

The WHILE condition is used to compare a variable to a range of values. The WHILE statement is used in place of a statement of the form (IF A>B AND IF A<C). WHILE statements simplify the control program by eliminating several IF-AND statements. It is often useful when modeling systems that must operate within a certain range of temperatures or pressures. Using a WHILE statement can allow you to incorporate an alarm or a shut down signal should the process reach unstable conditions, such as the limits of the range that the WHILE statement operates under. A simple example illustrating the use of the WHILE statement is shown below.

Example:

A tank that is initially empty needs to be filled with 1000 gallons of water 500 seconds after the process has been started-up. The water flow rate is exactly 1 gallon/second if V1 is completely open and V1 controls the flow of water into the tank.

Using a IF...THEN statement the program could be written as follows: IF t >500 and t <1501 THEN set V1 to open

ELSE set V1 to close.

The WHILE statement used to describe this relationship is as follows:

WHILE 500< t <1501 set V1 to open

ELSE set V1 to close.

It may not seem like much of a change between the two forms of the code. However, this is a very simple model. If modeling a process with multiple variables you could need many IF...THEN statements to write the code when a single WHILE condition could replace it.

Example:

V1 controls reactants entering a reactor that can only run safely if the temperature is under 500K.

The WHILE can be used to control the process as follows:

WHILE T <500 set V1 to open

ELSE set V1 to close. ALARM.

This example shows how a WHILE condition can be used as a safety measure to prevent a process from becoming unstable or unsafe.

In addition to lists of IF-THEN-ELSE-WHILE statements, control logic can be alternately represented by truth tables and state transition diagrams. Truth tables show all the possible states of a model governed by conditional statements and state transition diagrams represent truth tables graphically. Oftentimes, these are used in conjunction with booleans, variables that can only have two values, TRUE OR FALSE. A Boolean model or Boolean function follows the format of IF-THEN statements described here.

GO TO Statements

The GO TO statement helps to break out of current run to go to a different configuration. It can be an important operator in logical programming because a lot of common functions are accessed using GO TO. However, many programmers feel that GO TO statements should not be used in programming since it adds an extra and often unnecessary level of complex that can make the code unreadable and hard to analyze. Even though the GO TO operator has its downsides, it is still an important operator since it can make help to simplify basic functions. It can simplify code by allowing for a function, such as a fail safe, be referenced multiple times with out having to rewrite the function every time it is called.

The GO TO operator is also important because even advanced languages that do not have a GO TO function often have a different operator that functions in a similar manner but with limitations. For example, the C, C++ and java languages each have functions break and continue which are similar to the GO TO operator. Break is a function that allows the program to exit a loop before it reaches completion, while the continue function returns control to the loop without executing the code after the continue command. A function is a part of code within a larger program, which performs a specific task and is relatively independent of the remaining code. Some examples of functions are as follows:

- FUNCTION INITIALIZE: It runs at the beginning of a process to make sure all the valves and motor are in correct position. The operation of this function could be to close all valves, reset counters and timers, turn of motors and turn off heaters.
- FUNCTION PROGRAM: It is the main run of the process.
- FUNCTION FAIL SAFE: It runs only when an emergency situation arises. The operation of this function could be to open or close valves to stop the system, quench reactions via cooling, dilution, mixing or other method.
- FUNCTION SHUTDOWN: It is run at the end of the process in order to shutdown.
- FUNCTION IDLE: It is run to power down process.

All the functions mentioned above except FUNCTION IDLE are used in all chemical processes.

ALARM Statements

The ALARM statement is used to caution the operators in case a problem arises in the process. Alarms may not be sufficient danger to shut down the process, but requires outside attention. Some example when ALARM statements are used in the process are as follows:

- If the storage tank of a reactant is low, then ALARM.
- If the pressure of a reactor is low, then ALARM.
- If no flow is detected even when the valve is open, then ALARM.

- If the temperature of the reactor is low even after heating, then ALARM.
- If redundant sensors disagree, then ALARM.

In conclusion ALARM functions are very important in order to run a process safely.

Control Language in Industry

As stated before, these commands are all in pseudocode, and not specific to any programming language. Once the general structure of the controller is determined, the pseudocode can be coded into a specific programming language. Although there are many proprietary languages in industry, some popular ones are:

- Visual Basic
- C++
- Database programming (ex. Structured Query Language/SQL)
- Pascal
- Fortran

Pascal and Fortran are older languages that many newer languages are based on, but they are still used with some controllers, especially in older plants. Any experience with different computer languages is a definite plus in industry, and some chemical engineers make the transition into advanced controls designing, writing, and implementing code to make sure a plant keeps running smoothly.

Logical Functions in Microsoft Excel

Microsoft Excel has basic logical tools to help in constructing simple logical statements and if needed more complex logical systems. A list of the functions is shown below.

TRUE()..Returns the logical value, TRUE.

FALSE()...Returns the logical value, FALSE.

AND(logical_expression_A,B,C).........Returns TRUE if all the expressions are true.

OR(logical_expression_A,B,C)...........Returns TRUE if one of the expressions are true.

NOT(logical_expression)....................Returns the opposite of the expected logical value. If the expression is TRUE, it will return FALSE

IFERROR(value,value_if_error)............Returns the value unless there is an error in which it will return the value_if_error.

IF(logical_expression,value_if_true,value_if_false)..............Checks the validity of the expression and returns TRUE or FALSE likewise.

The IF() function will be most useful in logical programing. It is essentially an IF THEN or IF ELSE function returning one of two values based on the logical expression it is testing. Excel also allows logical functions within functions.

This allows for logical expressions involving more than one IF statement within itself for example. Use of these tools will be practical in quickly setting up control programs and other systems in Microsoft Excel.

Constructing a Logical Control Program

Understanding the conditional statements used in control logic is the first step in constructing a logical control program. The second step is developing a thorough understanding of the process to be controlled. Knowledge of the equipment, piping, and instrumentation, operating conditions, chemical compounds used, and safety concerns is necessary. Particularly it is important to know the measured and controlled variables.

For example, the pressure limits of a tank must be known in order to develop a control plan to ensure safety; ignoring this constraint could lead to explosion and injury. Once the necessary controls are known, one can develop a plan using the logical statements described previously. The third step is constructing a logical control program is understanding that there is not always a right answer, meaning there are many different ways to ensure the same desired outcome.

Worked out Examples 1, 2, and 3 demonstrate the construction of simple logical control programs. The more complex the situation, the longer the logical control plan becomes yet the process is still the same.

Determining Fail Safe Conditions

Fail safe is the practice of designing a system to default to safe conditions if anything or everything goes wrong. The goals of fail safe conditions are to:

- Protect plant personal
- Protect the local community around the plant
- Protect the environment
- Protect plant equipment

In order to establish safe conditions, fail safe programs must specify the desired positions of all valves and status of all motors and controlled equipment. For example, in an exothermic reactor, fail safe conditions would specify opening all cooling water valves, closing all feed valves, shutting off feed pump motors, turning on agitator motor, and open all vent valves.

Control programs frequently define fail safe conditions at the beginning of the program. These conditions are then activated using a GO TO command when process conditions exceed the maximum or fall below the minimum allowable values.

All processes must be evaluated for conditions that could cause hazards and fail safe procedure must be designed counteract the effects.

Modeling Case Studies

Surge Tank Model

Used to regulate fluid levels in systems, surge tanks act as standpipe or storage reservoirs that store and supply excess fluid. In a system that has experienced a surge of fluid, surge tanks can modify fluctuations in flow rate, composition, temperature, or pressure. Typically, these tanks (or "surge drums") are located downstream from closed aqueducts or feeders for water wheels. Depending upon its placement, a surge tank can reduce the pressure and volume of liquid, thereby reducing velocity. Therefore, a surge tank acts as a level and pressure control within the entire system.

Since the flow to the surge tank is unregulated and the fluid that is output from the surge tank is pumped out, the system can be labeled as unsteady-state [MIT], but the approach to an approximate solution (below) utilizes techniques commonly adhered to when solving similar steady-state problems.

The technology behind surge tanks has been used for decades, but researchers have had difficulty fully finding a solution due to the non-linear nature of the governing equations. Early approximations involving surge tanks used graphical and arithmetical means to propose a solution, but with the evolution of computerized solving techniques, complete solutions can be obtained. [Wiley InterScience].

Derivation of Ordinary Differential Equation

To accurately model a surge tank, mass and energy balances need to be considered across the tank. From these balances, we will be able to develop relationships for various characteristics of the surge tank

Diagram of Surge Tank System

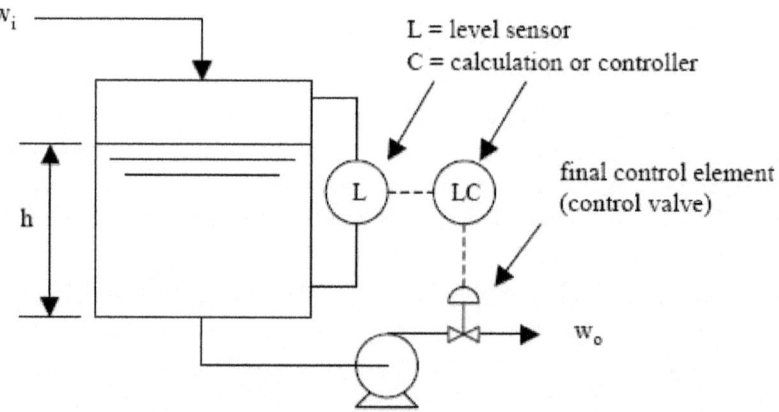

A surge tank relies on the level sensor to determine whether or not fluid stored in the tank should be removed. This regulated outflow is pumped out in

response to calculations made by a controller that ultimately opens and closes the control valve that releases the fluid from the tank.

Governing Equations of Surge Tank Model

A surge tank's components must be divided up and evaluated individually at first, then as a whole. First, the expression for the inlet stream must be obtained. The simplistic sine function will be used as the basis for the expression of a stream because it typically describes the tidal surge pattern of a low-viscous fluid, like water. The flowrate, w, will be given in units of kg h^{-1}.

$$w_i = a + bsin(c\pi t)$$

where a and b are constants determined by the specific problem circumstance.

A mass balance must now be performed on the tank as a system. Using the concept that *mass in = mass out,* and assuming that the tank is a perfect cylinder devoid of diversions, an expression can be derived:

rate mass in - rate mass out = accumulated fluid in tank

Rewriting,

$$w_i - w_o = \rho\frac{dV}{dt}$$

$$\frac{dh}{dt} = \frac{w_i - w_o}{\rho A}$$

$$\int_0^h dh = \int_0^t \frac{w_i - w_o}{\rho A}dt$$

$$h(t) - h_o = \frac{1}{\rho A}\int_0^t w_i - w_o dt$$

$$h(t) = h_o + \frac{1}{\rho A}\int_0^t w_i - w_o dt$$

Where, at time *t=0*, the amount of fluid in the surge tank is constant; thus $h(0) = h_0$.

Substituting the original equation for the inlet stream, w_i, into the expression for the height of the tank, $h(t)$, the governing equation for the height of the tank is obtained:

$$h(t) = h_o + \frac{1}{\rho A}\int_0^t [(a + bsin(c\pi t)) - w_0]dt$$

Integrating by parts,

$$h(t) = h_o + \frac{x}{\rho A}(1 - cos(c\pi t))$$

where$x$ is formed from the constants a and b during integration.

Secondary Uses of Surge Tanks

Surge tanks are most commonly used to protect systems from changes in fluid levels; they act as a reservoir that stores and supply excess fluid. In addition, the tanks shield the systems from dramatic changes in pressure, temperature, and concentration. They can also allow one unit to be shut down for maintenance without shutting down the entire plant.

Pressure

There can be moments of high pressure, called hammer shock, in a system when the liquid (incompressible) flow is stopped and started. The energy that liquids possess while traveling through pipes can be categorized as potential or kinetic.

If the liquid is stopped quickly, the momentum that the liquid carries must be redirected elsewhere. As a result, the pipes vibrate from the reactive force and weight of the shock waves within the liquid. In extreme cases, pipes can burst, joints can develop leaks, and valves and meters can be damaged.

The extreme amounts of pressure are dampened when the fluid enters the surge tanks. The surge tank acts as a buffer to the system, dispersing the pressure across a greater area. These tanks make it possible for a system to more safely execute their tasks.

Temperature

The temperature of a fluid can either be controlled or changed through the use of a surge tank. The surge tank allows for a rapid change in fluid temperature. This is exemplified by the process of pasteurization; the milk needs to be at a high temperature for just a short period of time, so it is exposed to the high temperature and then moved to the surge tank where it can be stored and cooled (see heated surge tank). A substance can enter the surge tank at room temperature, and it will instantaneously mix with the rest of the tank. Substances entering the tank will also subsequently rise to meet the high temperature and then exit the surge tank quickly thereafter.

Concentration

Concentration inside the surge tank is kept relatively constant, thus the fluid exiting the surge tank is the same as the fluid in the tank. This is favorable when there is a concentration gradient in the incoming fluid to the surge tank. The tank homogenizes the entering fluid, keeping the concentration of the reactants the same throughout the system, therefore eliminating any concentration gradient.

Heated Surge Tank Model

A surge tank is an additional safety or storage tank that provides additional product or material storage in case it becomes needed. Heat exchange can be added to surge tanks, which provides temperature control for the tank. Within a system these tanks can appear as distillation columns, reboilers, heated CSTR's, and heated storage.

They can increase production rates by allowing a batch of product to finish reacting while the initial tank is reloaded, provide constant system parameters during start up and shut down, or create additional storage space for product overflow or backup material.

Uses for Heated Surge Tanks:

- **Fuel surges caused by motion of a vehicle:** If fuel cannot be drawn from the primary tank, the engine resorts to a surge tank. The heat maintains the fuel's temperature.

- **Caramelization:** During the formation of caramel, the mixture must be maintained at a specific temperature for a predetermined amount of time. Once the ingredients are thoroughly dissolved, the mixture is transferred to a heated surge tank and maintained until the caramel has thickened and is ready to be drawn out.

- **Mixing of gases:** Bulk gas lines can be connected to a heated surge tank with a pressure sensor. The pressure sensor would control the temperature. By heating the gas when it first enters the tank, there is no risk of explosion later due to expansion.

- **Heated pools:** Surge tanks are used to catch and store displaced water from a pool. If the pool is heated, a heated surge tank should be used to maintain the temperature of the water.

- **De-aeration:** Heated surge tanks are often used with de-aerators. They heat the component that will enter the de-aerator, because if the component is not preheated, the de-aerator must wait until the component reaches the correct temperature. This could waste a lot of time and energy.

- **Chemical Baths:** Often in industry, things need to be treated with a chemical bath. The chemicals usually need to be at a certain temperature so that it will adhere to the object. A heated surge tank is perfect for this application.

- **Reboilers:** Liquids coming off of a distillation column can be reheated to enter the column again at a higher temperature to drive the separation process. Many industries use this tool to obtain a more efficient separation and produce a higher net profit.

- **Product or Material Backup:** Heated surge tanks can also be used as simple storage in two ways. First, a surge tank can be used excess product not yet sold or otherwise moved out of the production system. Second, heated surge tanks can serve as backup for chemical or fuel supplies to a production plant, such as outdoor gasoline tanks for a backup generator in case of power failure.

Basic Design for Heated Surge Tanks

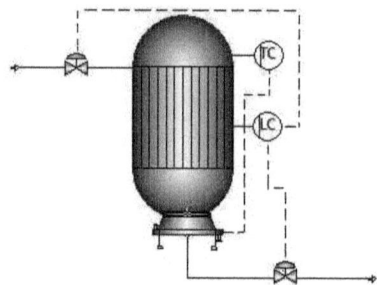

Above is a basic example of a heated surge tank. While surge tanks can have multiple inputs and outputs, for simplicity we have only included one of each here.

Connected to the tank is a temperature control, which controls the heater. Depending on the temperature of the fluid, this control will increase or decrease the heating to the tank. This will keep the fluid at the necessary temperature to meet the process requirements.

There is also a level control connected to the tank to indicate when the tank has neared maximum capacity. When this happens, the control will open the valve at the bottom of the tank, allowing the product to flow further down the process. The control can also slow or stop the flow coming through the input valve. These mechanisms will prevent the tank from overfilling. The position of the level control depends on the type of material in the process, the phase of the material, the type of level control, and the requirements of the system.

Useful Equations

The basic equations that govern heated surge tanks are shown below. First, a simple mass balance is done on the system. Second, the energy balance was simplified using the assumptions listed below. Most problems involving this type of tank can be described by these equations. Additional considerations may require additional variables and equations.

Assumptions:

1) The substance coming into the tank is uniform.
2) No reaction is taking place.
3) The tank is well mixed, which means the temperature profile is constant throughout the tank.

Mass Balance

Since there is no generation from reactions inside the heated surge tank, we obtain the rate of accumulation or level inside the tank by subtracting what is coming out from what is coming in.

$$(Rate\ of\ Accumulation) = (Flow\ In) - (Flow\ Out)$$

Energy Balance

The temperature change with respect to time is essential for the purpose of configuring a system to reach steady state. When turning a system on or off, there is a time period in which the system is in unsteady state. During this time, the system is difficult to model. In steady state, the system is easier to model because once steady state is reached the left hand term will become zero.

$$\frac{dT}{dt} = \frac{v\rho\, C_p(T_0 - T) + UA(T_C - T)}{V\rho\, C_p}$$

Case Study - Water Purification at IBM

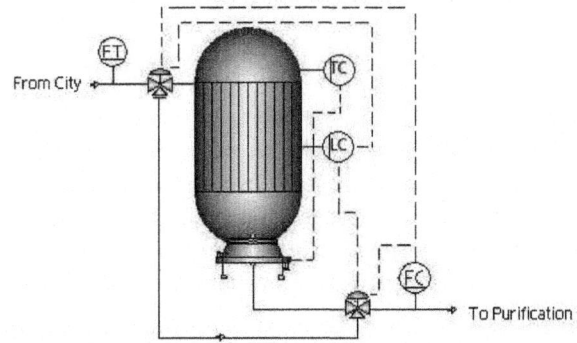

At IBM's manufacturing facility outside Burlington, Vermont, a heated surge tank is used in the de-ionized water system. In order to wash semi-conductor wafers in manufacturing, the water has to be about 1,000,000 times cleaner than the incoming city water. All of this purification is done on site.

The water comes in from the municipal water source at a constant flow rate, but manufacturing demand is not constant. In order to compensate for this, when the demand in manufacturing is low, a surge tank is used to store extra water for high demand periods. Because the large tank is located outside and the winter in Vermont is very cold, the tank is heated to prevent the water inside from freezing.

During normal operation of the system, the surge tank is bypassed. When a flow controller downstream has low demand, the inlet valve opens, letting water into the surge tank. A level controller monitors the tank to make sure it doesn't overfill and can shut off the inlet valve and let water out. A temperature controller controls the heater jacket to maintain the water around 50°C. When the demand for water increases, the flow controller near the outlet can shut off the inlet valve to the tank, and/or further open the outlet valve to access the extra water supply in the tank.

Chapter 5

BACTERIAL CHEMOSTAT MODEL

Bioreactors are used to grow, harvest, and maintain desired cells in a controlled manner. These cells grow and replicate in the presence of a suitable environment with media supplying the essential nutrients for growth. Cells grown in these bioreactors are collected in order to enzymatically catalyze the synthesis of valuable products or alter the existing structure of a substrate rendering it useful. Other bioreactors are used to grow and maintain various types of tissue cultures.

Process control systems must be used to optimize the product output while sustaining the delicate conditions required for life. These include, but are not limited to, temperature, oxygen levels (for aerobic processes), pH, substrate flowrate, and pressure. A bacterial chemostat is a specific type of bioreactor. One of the main benefits of a chemostat is that it is a continuous process (a CSTR), therefore the rate of bacterial growth can be maintained at steady state by controlling the volumetric feed rate. Bacterial chemostats have many applications, a few of which are listed below.

Applications:

Pharmaceuticals: Used to study a number of different bacteria, a specific example being analyzing how bacteria respond to different antibiotics. Bacteria are also used in the production of therapeutic proteins such as insulin for diabetics.

Manufacturing: Used to produce ethanol, the fermentation of sugar by bacteria takes place in a series of chemostats. Also, many different antibiotics are produced in chemostats.

Food Industry: Used in the production of fermented foods such as cheese.

Research: Used to collect data to be used in the creation of a mathematical model of growth for specific cells or organisms.

The following sections cover the information that is needed to evaluate bacterial chemostats.

Bacterial Chemostat Design

The bacterial chemostat is a continuous stirred-tank reactor (CSTR) used for the continuous production of microbial biomass.

Chemostat Setup

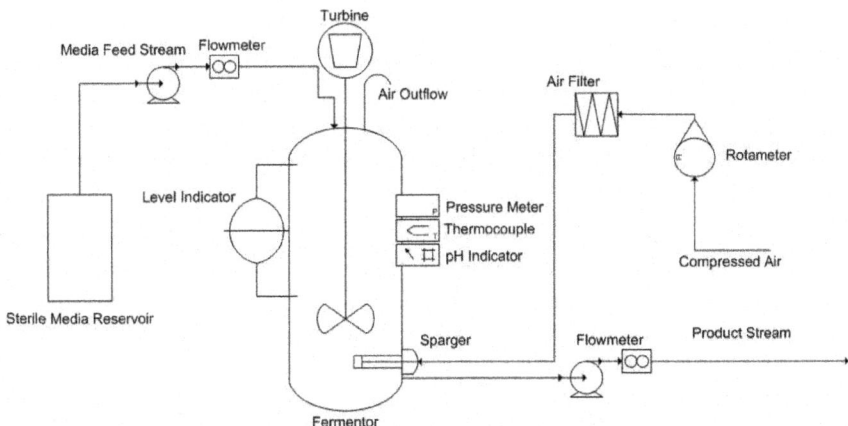

The chemostat setup consists of a sterile fresh nutrient reservoir connected to a growth chamber or reactor. Fresh medium containing nutrients essential for cell growth is pumped continuously to the chamber from the medium reservoir. The medium contains a specific concentration of growth-limiting nutrient (C_s), which allows for a maximum concentration of cells within the growth chamber.

Varying the concentration of this growth-limiting nutrient will, in turn, change the steady state concentration of cells (C_c). Another means of controlling the steady state cell concentration is manipulating the rate at which the medium flows into the growth chamber. The medium drips into culture through the air break to prevent bacteria from traveling upstream and contaminating the sterile medium reservoir.

The well-mixed contents of the vessel, consisting of unused nutrients, metabolic wastes, and bacteria, are removed from the vessel and monitored by a level indicator, in order to maintain a constant volume of fluid in the chemostat. This effluent flow can be controlled by either a pump or a port in the side of the reactor that allows for removal of the excess reaction liquid. In either case, the effluent stream needs to be capable of removing excess liquid faster than the feed stream can supply new medium in order to prevent the reactor from overflowing.

Temperature and pressure must also be controlled within the chemostat in order to maintain optimum conditions for cell growth. Using a jacketed CSTR for the growth chamber allows for easy temperature control. Some processes such as biological fermentation are quite exothermic, so cooling water is used to keep the temperature at its optimum level. As for the reactor pressure, it is controlled by an exit air stream that allows for the removal of excess gas.

For aerobic cultures, purified air is bubbled throughout the vessel's contents by a sparger. This ensures enough oxygen can dissolve into the reaction medium. For anaerobic processes, there generally is not a need for an air inlet, but there must be a gas outlet in order to prevent a build up in pressure within the reactor.

In order to prevent the reaction mixture from becoming too acidic (cell respiration causes the medium to become acidic) or too basic, which could hinder cell growth, a pH controller is needed in order to bring pH balance to the system.

The stirrer ensures that the contents of the vessel are well mixed. If the stirring speed is too high, it could damage the cells in culture, but if it is too low, gradients could build up in the system. Significant gradients of any kind (temperature, pH, concentration, *etc.* can be a detriment to cell production, and can prevent the reactor from reaching steady state operation.

Another concern in reactor design is fouling. Fouling is generally defined as the deposition and accumulation of unwanted materials on the submerged surfaces or surfaces in contact with fluid flow. When the deposited material is biological in nature, it is called biofouling. The fouling or biofouling in a system like this can cause a decrease in the efficiency of heat exchangers or decreased cross-sectional area in pipes.

Fouling on heat exchanger surfaces leads to the system not performing optimally, being outside the target range of temperature, or spending excess energy to maintain optimum temperature. Fouling in pipes leads to an increase in pressure drop, which can cause complications down the line. To minimize these effects, industrial chemostat reactors are commonly cylindrical, containing volumes of up to 1300 cubic meters, and are often constructed from stainless steel. The cylindrical shape and smooth stainless steel surface allow for easy cleaning.

Design Equations

The design equations for contiuous stirred-tank reactors (CSTRs) are applicable to chemostats. Balances have to be made on both the cells in culture and the medium (substrate).

Mass Balance

The mass balance on the microorganisms in a CSTR of constant volume is:

[Rate of accumulation of cells, g/s] = [Rate of cells entering, g/s] – [Rate of cells leaving, g/s] + [Net rate of generation of live cells, g/s]

The mass balance on the substrate in a CSTR of constant volume is:

[Rate of accumulation of substrate, g/s] = [Rate of substrate entering, g/s] – [Rate of substrate leaving, g/s] + [Net rate of consumption of substrate, g/s]

Assuming no cells are entering the reactor from the feed stream, the cell mass balance can be reworked in the following manner:

$$(Rate\ Accumulation\ Cells) = V\frac{dC_C}{dt}$$

$(Flow\ Entering) - (Flow\ Leaving) = 0 - \nu_0 C_C$

$(Rate\ Cell\ Generation) = V(r_g - r_d)$

Similarly, the substrate mass balance may be reworked in the following manner:

$(Rate\ Accumulation\ Substrate) = V\dfrac{dC_S}{dt}$

$(Flow\ Entering) - (Flow\ Leaving) = \nu_0 C_{S0} - \nu_0 C_S$

$(Rate\ Substrate\ Consumption) = V r_S$

Putting equations together gives the design equation for cells in a chemostat:

$$V\frac{dC_C}{dt} = 0 - \nu_0 C_C + V(r_g - r_d)$$

Similarly, equations together gives the design equation for substrate in a chemostat:

$$V\frac{dC_S}{dt} = \nu_0 C_{S0} - \nu_0 C_S + V(r_g - r_d)$$

Assumptions made about the CSTR include perfect mixing, constant density of the contents of the reactor, isothermal conditions, and a single, irreversible reaction.

Rate Laws

Many laws exist for the rate of new cell growth.

Monod Equation

The Monod equation is the most commonly used model for the growth rate response curve of bacteria.

$r_g = \mu C_c$

where r_g = cell growth rate

C_c = cell cencentration

μ = specific growth rate

The specific cell growth rate, μ, can be expressed as

$$\mu = \mu_{max}\frac{C_s}{K_s + C_s}$$

where μ_{max} = a maximum specific growth reaction rate

K_s = the Monod constant

C_s = substrate concentration

Tessier Equation and Moser Equation

Two additional equations are commonly used to describe cell growth rate. They are the Tessier and Moser Equations. These growth laws would be used when they are found to better fit experimental data, specifically at the beginning or end of fermentation.

Tessier Equation:

$$r_g = \mu_{max}[1 - exp(-\frac{C_s}{k})]C_c$$

Moser Equation:

$$r_g = \frac{\mu_{max}C_s}{1 + kC_s^{-\lambda}}$$

where λ and k are empirical constants determined by measured data.

Death Rate

The death rate of cells, r_d, takes into account natural death, k_d, and death from toxic by-product, k_t, where C_t is the concentration of toxic by-product.

$$r_d = (k_d + k_tC_t)C_c$$

Death Phase

The death phase of bacteria cell growth is where a decrease in live cell concentration occurs. This decline could be a result of a toxic by-product, harsh environments, or depletion of nutrients.

Stoichiometry

In order to model the amount of substrate and product being consumed/produced in following equations, yield coefficients are utilized. Y_{sc} and Y_{pc} are the yield coefficients for substrate-to-cells and product-to-cells, respectively. Yield cofficients have the units of g variable/g cells. Equation represents the depletion rate of substrate:

$$-r_s = Y_{sc}r_g + mC_c$$

Equation represents the rate of product formation:

$$r_p = Y_{pc}r_g$$

Control Factors

The growth and survival of bacteria depend on the close monitoring and control of many conditions within the chemostat such as the pH level, temperature, dissolved oxygen level, dilution rate, and agitation speed.

As expected with CSTRs, the pumps delivering the fresh medium and removing the effluent are controlled such that the fluid volume in the vessel remains constant.

pH Level

Different cells favor different pH environments. The operators need to determine an optimal pH and maintain the CSTR at it for efficient operation. Controlling the pH at a desired value during the process is extremely important because there is a tendency towards a lower pH associated with cell growth due to cell respiration (carbon dioxide is produced when cells respire and it forms carbonic acid which in turn causes a lower pH). Under extreme pH conditions, cells cannot grow properly, therefore appropriate action needs to be taken to restore the original pH (*i.e.* adding acid or base).

Temperature

Controlling the temperature is also crucial because cell growth can be significantly affected by environmental conditions. Choosing the appropriate temperature can maximize the cell growth rate as many of the enzymatic activates function the best at its optimal temperature due to the protein nature of enzymes.

Dilution Rate

One of the important features of the chemostat is that it allows the operator to control the cell growth rate. The most common way is controlling the dilution rate, although other methods such as controlling temperature, pH or oxygen transfer rate can be used. Dilution rate is simply defined as the volumetric flow rate of nutrient supplied to the reactor divided by the volume of the culture (unit: time-1). While using a chemostat, it is useful to keep in mind that the specific growth rate of bacteria equals the dilution rate at steady state. At this steady state, the temperature, pH, flow rate, and feed substrate concentration will all remain stable. Similarly, the number of cells in the reactor, as well as the concentration of reactant and product in the effluent stream will remain constant.

Negative consequences can occur if the dilution rate exceeds the specific growth rate. As can be seen in Equation below, when the dilution rate is greater than the specific growth rate ($D > \mu$), the dC_C/dt term becomes negative.

$$\frac{dC_C}{dt} = (\mu - D)C_C$$

This shows that the concentration of cells in the reactor will decrease and eventually become zero. This is called wash-out, where cells can no longer maintain themselves in the reactor. Equation represents the dilution rate at which wash-out will occur.

$$D_{max} = \frac{\mu_{max}C_{s0}}{K_s + C_{s0}}$$

In general, increasing the dilution rate will increase the growth of cells. However, the dilution rate still needs to be controlled relative to the specific growth rate to prevent wash-out. The dilution rate should be regulated so as to maximize the cell production rate. Figure below shows how the dilution rate affects cell production rate(DC_C), cell concentration (C_C), and substrate concentration (C_S).

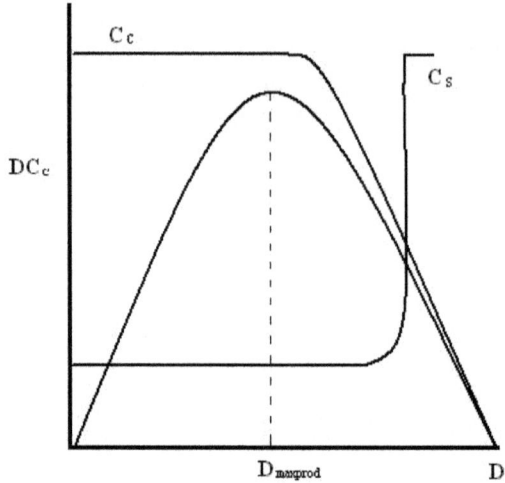

Fig. : Cell concentration, cell production, and substrate concentration as a function of dilution rate

Initially, the rate of cell production increases as dilution rate increases. When $D_{maxprod}$ is reached, the rate of cell production is at a maximum. This is the point where cells will not grow any faster. $D = \mu$ (dilution rate = specific growth rate) is also established at this point, where the steady-state equilibrium is reached.

The concentration of cells (C_C) starts to decrease once the dilution rate exceeds the $D_{maxprod}$. The cell concentration will continue to decrease until it reaches a point where all cells are washed out. At this stage, there will be a steep increase in substrate concentration because fewer and fewer cells are present to consume the substrate.

Oxygen Transfer Rate

Since oxygen is an essential nutrient for all aerobic growth, maintaining an adequate supply of oxygen during aerobic processes is crucial. Therefore, in order to maximize the cell growth, optimization of oxygen transfer between the air bubbles and the cells becomes extremely important. The oxygen transfer rate (OTR) tells us how much oxygen is consumed per unit time when given concentrations of cells are cultured in the bioreactor. This relationship is expressed in Equation below.

$$\text{Oxygen Transfer Rate (OTR)} = Q_{O2}C_C$$

Where C_C is simply the concentration of cell in the reactor and Q_{O2} is the microbial respiration rate or specific oxygen uptake rate. The chemostat is a very convenient tool to study the growth of specific cells because it allows the operators to control the amount of oxygen supplied to the reactor. Therefore it is essential that the oxygen level be maintained at an appropriate level because the cell growth can be seriously limited if inadequate oxygen is supplied.

Agitation Speed

A stirrer, usually automated and powered with a motor, mixes the contents of the chemostat to provide a homogeneous suspension. This enables individual cells in the culture to come into contact with the growth-limiting nutrient and to achieve optimal distribution of oxygen when aerobic cultures are present. Faster, more rigorous stirring expedites cell growth. Stirring may also be required to break agglutinations of bacterial cells that may form.

CSTR Heat Exchange Model

A CSTR (Continuous Stirred-Tank Reactor) is a chemical reaction vessel in which an impeller continuously stirs the contents ensuring proper mixing of the reagents to achieve a specific output. Useful in most all chemical processes, it is a cornerstone to the Chemical Engineering toolkit. Proper knowledge of how to manipulate the equations for control of the CSTR are tantamount to the successful operation and production of desired products.

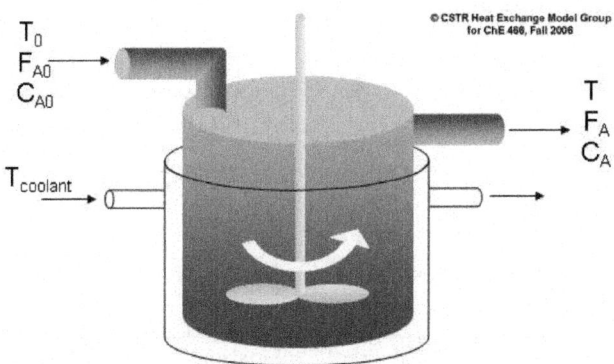

Fig. : CSTR with Heat Exchange.

The purpose of the model *dynamic conditions* within a CSTR for different process conditions. Simplicity within the model is used as the focus is to understand the dynamic control process.

Assumptions

We have made the following assumptions to explain CSTR with heat exchange modeling.

Perfect mixing

• The agitator within the CSTR will create an environment of perfect mixing within the vessel. Exit stream will have same concentration and temperature as reactor fluid.

Single, 1st order reaction

- To avoid confusion, complex kinetics are not considered in the following modeling.

$$-r_a = kC_A$$

Parameters specified

- We assume that the necessary parameters to solve the problem have been specified.

Volume specified

- In a control environment, the size of the vessel is usually already specified.

Constant Properties

- For this model, we have made the assumption that the properties of the species we are trying to model will remain constant over the temperature range at which we will be looking at. It is important to make this assumption, as otherwise we will be dealing with a much more complex problem that has a varying heat capacity, heat of reaction, *etc.*

ODE Modeling in Excel

To setup the model, the mass and energy balances need to be considered across the reactor. From these energy balances, we will be able to develop relationships for the temperature of the reactor and the concentration of the limiting reactant inside of it.

Variable Definitions

The following table gives a summary of all of the variables that we used in our mathematical formulas.

Symbol Explanations			
Symbol	**Meaning**	**Symbol**	**Meaning**
V	Volume of Reactor	ρ	Density of Stream
m	Mass Flow Rate	C_{A0}	Original Concentration
C_A	Current Concentration	T_0	Original Temperature
T	Current Temperature	T_C	Coolant Temperature
ΔH_{rxn}	Heat of Reaction	ΔC_p	Overall change in Heat Capacity
k_0	Rate Law Constant	E	Activation Energy
R	Ideal Gas Constant	UA	Overall Heat Transfer Coefficient

Mass Balance

From our energy and material balances coursework, we know that the general equation for a mass balance in any system is as follows:

$$(Rate\ Accumulation) = (Flow\ In) - (Flow\ Out) - (Rate\ Generation)$$

$$(Rate\ Accumulation) = V\frac{dC_A}{dt}$$

$$(Flow\ In) - (Flow\ Out) = \frac{m}{\rho}(C_{A0} - C_A)$$

$$(Rate\ Generation) = V(-r_a) = VC_A k_0 e^{-E/RT}$$

In the case of a CSTR, we know that the rate of accumulation will be equal to $V\frac{dC_A}{dt}$.

This comes from the fact that the overall number of moles in the CSTR is VC_A, so the accumulation of moles will just be the differential of this. Since V is a constant, this can be pulled out of the differential, and we are left with our earlier result. We also know that the flow of moles in versus the flow of moles out is equal to

$$\frac{m}{\rho}(C_{A0} - C_A),$$

which is the mass flow rate, divided by the density of the flow, and then multiplied by the difference in the concentration of moles in the feed stream and the product stream. Finally, we can determine what the rate of generation of moles in the system is by using the Arrhenius Equation. This will give us the rate of generation equal to

$$VC_A k_0 e^{-E/RT}.$$

Combining all of these equations and then solving for

$$\frac{dC_A}{dt},$$

we get that:

$$\frac{dC_A}{dt} = \frac{m}{\rho V}(C_{A0} - C_A) - k_0 C_A e^{-E/RT}$$

Energy Balance

From our thermodynamics coursework, we know that the general equation for an energy balance in any system is as follows:

$$(Rate\ Energy Accumulation) = (Heat\ In)-(Heat\ Out)+(Rate\ Heat\ Generation)+(Heat\ Transfer)$$

$$(Rate\ Energy Accumulation) = V\rho C_p \frac{dT}{dt}$$

$$(Heat\ In) - (Heat\ Out) = m\Delta C_p(T_0 - T)$$

$$(Rate\ Heat\ Generation) = -V\Delta H_{rxn}(-r_a) = -V\Delta H_{rxn}k_0 C_A e^{-E/RT}$$

$$(Heat\ Transfer) = UA(T_C - T)$$

In the case of a CSTR, we know that the rate of energy accumulation within the reactor will be equal to

$$V\rho C_p \frac{dT}{dt}.$$

This equation is basically the total number of moles (mass actually) in the reactor multiplied by the heat capacity and the change in temperature. We also know that the heat generated by this reaction is

$$-V\Delta H_{rxn}k_0 C_A e^{-E/RT},$$

which is the rate of mass generation ($-Vr_a$) times the specific heat of reaction (ΔH_{rxn}). The overall rate of heat transfer into and out of the system is given by

$$m\Delta C_p(T_0 - T).$$

This equation is the flow rate multiplied by the heat capacity and the temperature difference, which gives us the total amount of heat flow for the system. Finally, the amount of heat transferred into the system is given by

$$UA(T_C - T).$$

Combining all of these equations and solving the energy balance for $\frac{dT}{dt}$, we get that:

$$\frac{dT}{dt} = \frac{m\Delta C_p(T_0 - T) - V\Delta H_{rxn}C_A k_0 e^{-E/RT} + UA(T_C - T)}{V\rho\Delta C_p}$$

In a realistic situation in which many chemical processed deal with multiple reactions and heat effects slight changes to the modeled equation must be done. The diagram below evaluates the heat exchanger under heat effects in which there is an inlet and outlet temperature that is accounted for in the enthalpy term in the newly modeled equation.

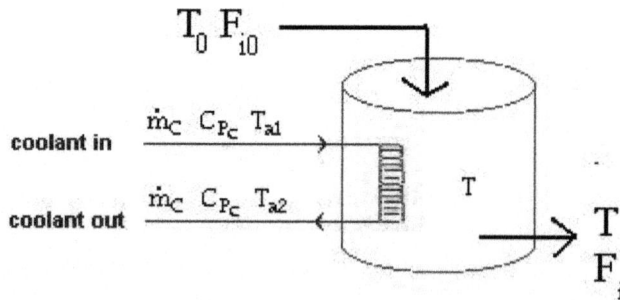

To model a heat exchanger that accounts for multiple reactions simply take the deltaHrxn and deltaCp term and add the greek letter sigma for summation in front of the terms. When considering a case with multiple reactions and heat effects, the enthalpy and heat capacity of each reaction must be implemented in the energy balance, hence i and j represents the individual reaction species. The equation now looks something like this:

$$\frac{dT}{dt} = \frac{m\sum \Delta C_{pij}(T_0 - T) - V\sum \Delta H_{rxnij}C_A k_0 e^{-E/RT} + UA(T_C - T)}{V\rho \sum \Delta C_{pij}}$$

Euler's Method

In order to model an ODE in Excel, we will be using Euler's Method. The reason why we chose Euler's Method over other methods (such as Runge-Kutta) was due to the simplicity associated with Euler's. Trying to model with methods such as Runge-Kuttaare much more difficult in Excel, while Euler's method is quite simple. We also checked the accuracy of our results using Euler's Method by comparing our answers with that of Polymath, a highly accurate ODE modeling program.

We found that our answers were the same as that of Polymath, so we are quite comfortable using Euler's Method for our model. Here are the full details on Euler's Method. For this model, we will primarily be interested in the change in concentration of a reactant and the temperature of the reactor. These are the two differential equations that we were able to obtain from the mass and energy balances in the previous section. The following is an application of Euler's method for the CSTR on the change in concentration:

$$C_{Ai} = C_{A(i-1)} + \frac{dC_A}{dt}\Delta t$$

This same application can be made to the change in temperature with respect to time:

$$T_i = T_{i-1} + \frac{dT}{dt}\Delta t$$

Assuming that all values in the ODE's remain constant except for T and C_A, the new value is then found by taking the pervious value and adding the differential change multiplied by the time step.

List of Equations

The following are a summary list of all of the equations to be used when modeling CSTR with a heat exchange.

$$k = k_0 e^{-E/RT}$$

$$\frac{dC_A}{dt} = \frac{m}{\rho V}(C_{A0} - C_A) - k_0 C_A e^{-E/RT}$$

$$\frac{dT}{dt} = \frac{m\Delta C_p(T_0 - T) - V\Delta H_{rxn}C_A k_0 e^{-E/RT} + UA(T_C - T)}{V\rho C_p}$$

$$C_{Ai} = C_{A(i-1)} + \frac{dC_A}{dt}\Delta t$$

$$T_i = T_{i-1} + \frac{dT}{dt}\Delta t$$

Combining Using Excel

To model the CSTR using Excel you use Euler's Method, Energy Balance and Mass Balance together to solve for the concentration at time t.

	A	B	C	D	E	F	G	H
10	m		4 kg/s	Mass feed rate and product rate				
11	k_0	1.97E+20 s^{-1}		Rate constant				
12	UA		7000 J/s	Heat Transfer Coefficient and Area				
13	T_0		400 K	Feed temperature				
14	$T_{0,2}$		400 K	Change Initial Feed Temperature (leave the same if this				
15	t for $T_{0,2}$		0 s	Time when Initial Feed Temperature Changes				
16	T_c		350 K	Coolant temperature				
17	$T_{c,2}$		350 K	Change Initial Coolant Concentration (leave the same if				
18	t for $T_{c,2}$		0 s	Time when Initial Coolant Temperature Changes				
19	V		3000 L	Reactor volume				
20	ρ		1 kg/L	Density of reactor feed and product (assumed constant)				
21	ΔH_{Rx}		-200 J/mol	Heat of reaction				
22								
23								
24	t (s)	T_0	T (K)	T_c	C_{A0}	C_A	dC$_A$/dt	dT/dt
25	0	400	400	350	9	9	0.00E+00	0
26	1	400	400	350	9	9	-3.72E-01	-0.04227
27	2	400	399.9577	350	9	8.627994	-3.54E-01	-0.04556
28	3	400	399.9122	350	9	8.273736	-3.37E-01	-0.04869
29	4	400	399.8635	350	9	7.936446	-3.21E-01	-0.05167
30	5	400	399.8118	350	9	7.615362	-3.06E-01	-0.05448

Screenshot of Excel Model for CSTR with Heat Exchange

How to Use Our Model

In order to help facilitate understanding of this process, we have developed an Excel spreadsheet specifically for looking at the changes in concentration and temperature given some change in the input to the CSTR system.

An example of a change to the system could be that the temperature of the feed stream has dropped by a given number of degrees, or that the rate at which the feed stream is being delivered has changed by some amount. By using our spreadsheet, you will be able to easily plug in your given parameters, and look at the trend of the concentration and temperature over a wide time interval.

The way in which this spreadsheet works is quite simple. Boxes are provided for you to input all of the given information for your CSTR problem. Various columns containing values for the temperature, concentration, *etc.*, with respect to time have also been provided. There are then more columns that contain the values for the various differential equations from above. With the time derivative in hand, we are then able to predict the value of the temperature or concentration at the next given time interval.

Our easy-to-use Excel model is given here: CSTR Modeling Template. In our model, you will find a column of unknowns that must be specified in order to solve for the optimal conversion temperature and optimal concentration of A. There are then two cells that will display the optimal temperature and concentration. Graphs are also provided to look at the change in temperature and concentration over time. Most of the variables in the model are self-explanatory.

One important feature of our model is the option to have a change in the temperature of the feed stream or the concentration of A after a given time t. You do not need to input a value for these cells if there is no change in the feed; it just provides a convenient way to look at the change of temperature and concentration of A. You are also provided with a cell for the time step, Δt. Depending on what size time step you choose, you may need to choose a larger value if your graphs do not reach steady state. If this is the case, the output cells will tell you to increase the time step.

APPLICATIONS TO PROCESS CONTROL

The model created can account for many different situations. However, in process control, it is very important for the following to be considered.

Exothermic reaction - In industry, the exothermic reaction is typically of more importance to controls as there is a safety factor involved. (*i.e.* explosive reactor conditions)

Volume constant - For CSTRs, liquids are mostly used.

The model developed can assist in controlling an actual process in many ways including the following, but not limited to:

- Maintaining product concentration despite changes operating conditions
- Predicting and preventing explosive reactor conditions
- Flow rate optimization
- Input concentration optimization
- Operating temperature optimization
- Coolant temperature optimization

This model is very useful for many simple reactor situations and will aid in the understanding of a process and how dynamic situations will affect operation.

In real-life applications, the reaction may not be first order or irreversible or involve multiple reactions. In all these cases, we modify the Energy and Mass Balance with additional rate law considerations.

Distillation Model

Distillation is a commonly employed separation technique bases on difference in volatilities. The modern form of distillation as is known today may be credited to early Arab alchemist, Jabir ibnHayyan and the development of one of his inventions, the alembic.

The distillation apparatus is commonly referred to as a still and consists of a minimum of a reboiler in which mixture to be separated is heated, a condenser in which the vapor components are cooled back to liquid form, and a receiver in which the concentrated liquid component fractions are collected.

Ideally, distillation is governed by the principles of Raoult's Law and Dalton's Law. Dalton's Law states that for a mixture, the total vapor pressure is equal to the sum of the individual vapor pressures of the pure components which comprise this mixture. The relationship giving the vapor pressure of a volatile component in a mixture, P_A, is Raoult's Law and is governed by the following equation:

$$P_A = X_A P_A^{\circ}$$

Where X_A is the mole fraction of component A in the mixture and P_A° is the vapor pressure of pure component A. This ideal model is based on a binary mixture of benzene and toluene but for other mixtures severe deviations from Raoult's Law may be observed due to molecular interactions. For these aforementioned mixtures where the components are not similar the only accurate alternative is obtaining vapor-liquid equilibrium by measurement.

In simple distillation, two liquids with differing boiling points are separated by immediately passing the vapors from the reboiler to a condensing column which condenses the vapor components. As a result the distillate is not usually pure but its composition may be determined by Raoult's Law at the temperature and pressure at which the vapors were boiled off.

Consequently, simple distillation is usually used to separate binary mixtures where the boiling temperatures of the individual components are usually significantly different or to separate volatile liquids from non-volatile solids.

A reference of terms used in distillation reference are included at the end of this article.

DISTILLATION CONTROL

Distillation columns comprise an enormous amount of the separation processes of chemical industries. Because of their wide range of uses in these industries and because their proper operation contributes to product quality, production rates and other capital costs, it is clear that their optimization and control is of great importance to the chemical engineer. Distillation control becomes problematic because of the wide variety of thermodynamic factors stemming from the separation process.

For example:

- Separations deviate from linearity of equations as product purity increases.
- Coupling of process variables occurs when compositions are controlled.
- Disturbances occur due to feed and flow agitation
- Efficiency changes in trays lead to non-steady state behaviour.

In order to improve upon distillation control you must be able to characterize these potential problems and realize when they occur because they lead to
dynamic behaviour of the column.

Of key importance to control is the maintenance of material and energy balances and their due effects on the column. Shown below is a schematic of a simple
binary distillation column. Using the material balance formulas

$$D/F = (z-x)/(y-x),$$

where z, x, and y are the feed, bottoms and distillate concentrations respectively,
you find that as D (Distillate) increases, its purity decreases. This leads to the idea
that purity level varies indirectly with the flow rate of that product. Energy input
is also key because it determines the vapor flow rate (V) up the column which
has direct effects on the L/D ratio (reflux ratio) and hence relates to an increase
in the amount of separation occurring. To summarize, energy input determines
the amount of separation, while material flow relates the ratio of separation in
the products.

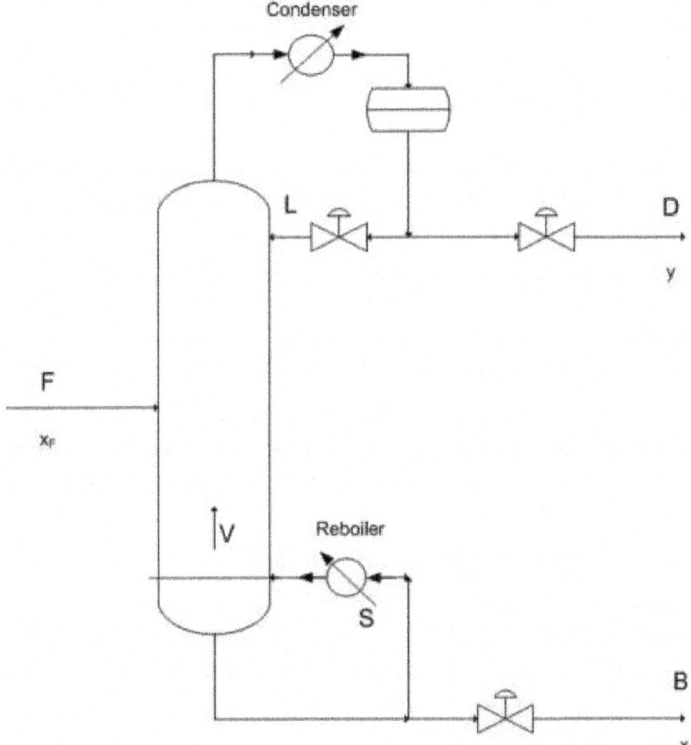

Vapor liquid dynamics within the column also contributes to the theory
behind process control because of a few important relations. Changing V (by
changing the reboiler energy), causes an extremely rapid response in the overhead composition, while changing the reflux ratio requires a longer response to
its effect on the reboiler.

In lower pressure columns, a phenomena known as entrainment or flooding occurs in which liquid is blown up into trays instead of dropping down into trays. This significantly decreases separation efficiency and therefore less product gain occurs. Using a packed column in these low pressure applications provides greater efficiency over tray columns and also allows for the faster accomplishment of a steady state profile. Controlling the occurrence of entrainment in either case is another crucial aspect which should be recognized when designing control systems for columns.

Regulatory Controls

For the distillation process it is imperative that regulatory controls such as level, flow, and pressure controllers are functioning properly to further ensure the effectiveness of the product composition controllers.

In terms of regulatory control, *level controls* are used to maintain specified levels in the reboiler, accumulator, and in the case of a distillation column with two columns due to high tray numbers for a single column, also maintain the level in an intermediate accumulator. Inept use of level controls may lead to problems elsewhere in the distillation process.

For example, poor level control on the accumulator and reboiler may lead to problems with composition control for material balance control configurations. Also if the reboiler duty is maintained by one of these level controllers and the controller causes oscillation in the reboiler, consequently cycling may also occur in column pressure.

Flow controllers are used to manipulate and maintain desired flow rates of the reflux, distillate and bottoms products, and the heating medium employed in the reboiler. The setpoints of this specific type of controller is determined by the various composition and level controllers in the process.

Pressure controls are located to the top of the distillation column in the distillate vicinity. Here the column overhead pressure, caused by the accumulation of components in the vapor phase, acts as an integrator causing a change in the level of the accumulator. This pressure may be controlled using a variety of methods.

Pressure Control through Condenser Operation

1) Maximize cooling water flow rate to the condenser (operate at minimum column pressure)
2) Adjust the rate of condensation of the overhead (for example by adjusting the flow rate of refrigerant to the condenser)
3) Adjust level of liquid in the condenser (changes heat transfer area)

Pressure Control through Accumulator Operation

1) Purge vapor from the overhead accumulator
2) Directly changing amount of material phase (for example by pumping in inerts)

Composition and Constraint Control

Because distillation requires a desired product concentration or flowrate, constraint control is used to ensure the desired operation conditions by having setpoints designated for the requirements of the system. The constraint is usually a concentration and the control of this concentration can vary depending on the application. In almost all industry applications only one product concentration is controlled, while the other is allowed to vary.

This is known as Single Composition control and is much easier to achieve and maintain than the Dual composition control which specifies both product concentrations. The advantage to Dual Control however being increased energy efficiency because of increased separation. P&ID of common control placement used for composition control are shown below. Control lines are not included in this diagram because of the various number of control possibilites for a system.

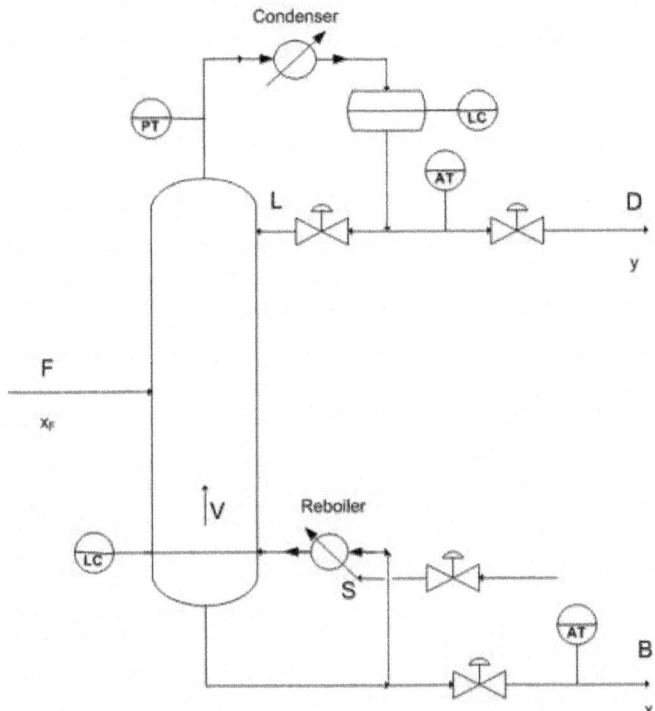

Setting constraints on a column allows for proper control of the product as well as points to issues of safety and maintenance. The most common constraint controls are in the maximum reboiler and condenser duty which results from a number of variables including fouling [[1]], improper valve sizing, and excessive increases in feed. Other common constraints are flooding and weeping points [[2]] which indicate incorrect L/V ratios and can be compensated for by adjusting the pressure drop across the column.

The most common adjustments for constraint control involve changing the reboiler duty to satisfy the constraints and almost always follow with subsequent fall back mechanisms to ensure product quality.

1. Ensuring single composition control
2. Reducing feed rate
3. Increasing product purity setpoints

Effective Distillation Control

Before doing an in depth analysis of distillation control, it is of paramount importance that the following basics are attended to.

- Firstly, ensure that regulatory controls are indeed functioning congruously.

- For changes in reflux temperature, employ the use of reflux controls.

- Make sure to check and evaluate analyzer deadtime, accuracy, and reliability. This is required to account for the lagtime from the product composition analyzer when using feedback control to control feed flow, reflux ratio and reboiler power. Refer to [[3]] for deadtime. To select proper analyzer refer to [[4]]

- Ensure that any thermistors or RTD's employed to measure tray temperatures for composition inference are fully operational and correctly situated. Care should be taken here to ensure that pressure corrected temperatures are used. Refer to the following for temperature sensors [[5]]

- When streams such as D, B, L, and V are used as manipulated variables for composition control, they should be changed with respect to the measured feed rate when column feed rate changes are a common disturbance.

Summary of Distillation Control Methods	
Control Parameter	**Example of Control Method**
Distillate flow rate (D)	Flow controller (Setpoint controlled by accumulator level)
Bottoms flow rate (W)	Flow controller (Setpoint controlled by level in bottom of column)
Reflux flow rate (L_D)	Flow controller (Setpoint controlled by top tray temperature)
Reboiler Steam flow rate	Flow controller (Setpoint controlled by bottoms composition analyzer)
Feed flow rate	Flow controller (Setpoint manually entered based on unit operations)
Distillate purity (x_D)	Reflux flow controller (Setpoint controlled by top tray temperature)
Bottoms purity (x_W)	Steam flow controller (Setpoint controlled by online analyzer)
Column pressure	Purge flow controller (Setpoint controlled by column pressure)

Control Problems and Disturbances

Feed Composition & Feed Flow Upsets

In order to properly determine the ultimate product purity and flow you must consider the impact of disturbances in the column system. The most significant, however most easily remedial of these disturbances is a feed composition upset, in which there is a change in the feed composition resulting in a major disturbance in the product composition.

Hence, configuration of a distillation column must consider the control regulations of such a feed upset. Feed flow upsets are regulated using ratio control of L/F, D/F, V/F and B/F through level and reflux ratio sensors.

Feed and Reflux Enthalpy Disturbances

Feed Enthalpy upsets become an issue for columns operating at low reflux ratios and cause a large deviation from expected product concentrations due to the changes in vapor and liquid flowrates in the column. The usual compensation for this is using a feed heat exchanger to control the proper enthalpy to the column. In most cases, the feed is preheated before entering the column.

By adjusting the duty of the preheater (*i.e.* decreasing or increasing the heating medium flow rate), a constant vapor/liquid ratio in the feed can be maintained. Rapid changes to the external conditions of a column, (especially large temperature deviations in rainstorms) can cause a subcooled reflux, changing the composition of the products. Reflux control can amend this properly.

Steam and Column Pressure Upsets

The most severe disturbance occurring in a distillation column occurs when there is a loss of steam pressure in the reboiler. A sharp drop in steam pressure results in a drop in reboiler effectiveness and therefore a huge increase in the impurity of the product. This can be avoided using an override control loop for this particular occurrence.

Due to the effect of pressure on the relative volatility of the components in the system, a disturbance in column pressure leads to altered product quality. This can be effectively maintained by a composition controller to compensate for these pressure differences.

ODE Modeling of a Distillation Column

There are two methods used for modeling a distillation column: dynamic modeling and steady state modeling. The key difference between the models is that dynamic modeling is used to monitor changes in the distillation column as a function of time, while the steady state model looks at a given set of conditions at one particular time (*i.e.* when the column is at steady state).

Dynamic Model

Dynamic modeling of a distillation column may be used for a variety of different reasons: monitoring variations in the column as a result of feed changes, to predict the effects of tray fouling, and to predict when flooding with occur. The dynamic model allows the user to improve distillation control by being able to deal with disturbances that cause upsets to the column's normal operation.

The algorithm for developing a dynamic distillation column model is as follows:

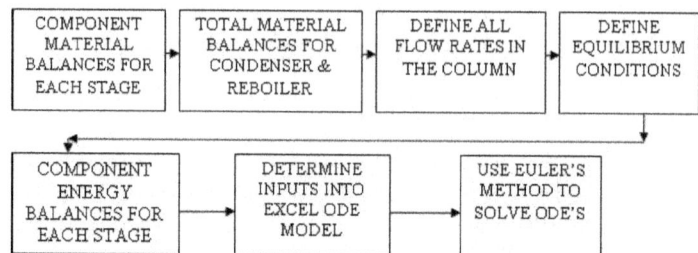

Step 1) Write component material balance for each stage in the column.

accumulation = liquid entering stage i + vapor entering stage i − liquid leaving stage i − vapor leaving stage i

Component material balance for all stages, except the feed tray, overhead condenser, and reboiler:

$$\frac{dM_i x_i}{dt} = L_{i-1} x_{i-1} + V_{i+1} y_{i+1} - L_i x_i - V_i y_i$$

Assumption: For simplicity, accumulation in the each stage is constant; $\dfrac{dM_i}{dt} = 0$

Simplified component material balance for each stage(only composition changes with time):

$$M_i \frac{dx_i}{dt} = L_{i-1} x_{i-1} + V_{i+1} y_{i+1} - L_i x_i - V_i y_i$$

The following are examples of equations used in the Excel Interactive ODE Distillation Column Model, which are provided to help the user understand how the model works.

ODE used to solve for the liquid composition leaving tray 2(rectifying section):

$$\frac{dx_2}{dt} = \frac{1}{M_2} \left[L_1 x_1 + V_3 y_3 - L_2 x_2 - V_2 y_2 \right]$$

ODE used to solve for the liquid composition leaving tray 5(stripping section):

$$\frac{dx_5}{dt} = \frac{1}{M_5} \left[L_4 x_4 + V_6 y_6 - L_5 x_5 - V_5 y_5 \right]$$

Overhead condenser component balance:

$$\frac{dx_D}{dt} = \frac{1}{M_D} \left[V_1 \left(y_1 - x_D \right) \right]$$

Feed tray component balance:

$$\frac{dx_3}{dt} = \frac{1}{M_3} [L_2 x_2 + V_4 y_4 - L_3 x_3 - V_3 y_3]$$

Reboiler component balance:

$$\frac{dx_W}{dt} = \frac{1}{M_W} [L_6 x_6 - W x_W - V_7 y_7]$$

Step 2) Write total material balances around condenser and reboiler

Condenser material balance:

Assumption 1: Total condenser (all vapor from the top of the column is condensed into a liquid).

Assumption 2: Overhead accumulator liquid level remains constant.

$$D = [V_1 + L_D]$$

Reboiler material balance:

$$W = [F - D]$$

For these equations to work, the user must specify:

- reflux flow rate (mol/min)
- bottoms flow rate (mol/min).

Step 3) Define all flow rates

Vapor Leaving Feed Stage:

$$V_3 = V_4 + F(1 - q_F)$$

Liquid Leaving Feed Stage:

$$L_3 = L_2 + F(q_F)$$

Vapor flow rates in stripping section:

Assumption: Equimolal overflow for vapor in stripping section

$$V_4 = V_5 = V_6 = (V_7)$$

Vapor flow rates in rectifying section:

Assumption: Equimolal overflow for vapor in rectifying section

$$V_1 = V_2 = (V_3)$$

Liquid flow rates in rectifying section:

Assumption: Equimolal overflow for liquid in rectifying section

$$L_2 = L_1 = (L_D)$$

Liquid flow rates in stripping section:

Assumption: Equimolal overflow for liquid in stripping section

$$L_6 = L_5 = L_4 = (L_3)$$

Step 4) Define equilibrium conditions

The binary system used in the Excel ODE model is a benzene-toluene system. The equilibrium data for this system was put in the model and the relative volatilities were calculated for various equilibrium compostions.

Relative Volatility (from equilibrium data):

$$\alpha = \frac{y_{benzene}\,x_{toluence}}{x_{benzene}\,y_{toluence}}$$

where α is defined as the relative volatility of the two components in the system.

These relative volatilities were plotted against temperature and linear regression was used to fit the data.

Relative volativity as a function of temperature:

$$\alpha = [-0.009\ T + 3.3157]$$

This equation models how the separation changes on each tray as a function of tray temperature, which decreases up the column.

Equilibrium Vapor Composition for each stage:

Assumption: Trays in the Column are 100% efficient (vapor and liquid leaving a tray are in equilibrium)

$$y_i = \frac{\alpha x_i}{1 + (\alpha - 1)x_i}$$

Replacing alpha with the temperature dependent equation shows how tray temperature affects the amount of benzene in the vapor leaving each tray.

Step 5) Write component energy balances for each stage.

The ODE energy balances are essential for the dynamic model to run properly. Mass transfer occurs within the column because the temperature varies from the top of the column to the bottom thereby allowing separation of the components in the system.

Since the energy input into the column is added in the reboiler, the reboiler ODE is the first equation entered in the model. In our model, this is given as:

$$\frac{dT_7}{dt} = \frac{1}{M_W}[L_6 x_6 - W\,x_W][T_6 - T_7] + \frac{q_r}{M_W c_p}$$

The next step is adding energy balances for each subsequent stage in the distillation column. The only stage in the column which has a slightly different energy ODE is the feed stage, given by:

$$\frac{dT_3}{dt} = \frac{1}{M_3}\big[[L_2 x_2][T_2 - T_3] + [V_4 y_4][T_4 - T_3] - [L_3 x_3][T_2 - T_3] + [V_3 y_3][T_4 - T_3] + [Fx_{feed}][T_{feed} - T_3]\big]$$

The last energy balance is around the condenser.

Assumption: Reflux return temperature is constant (overhead condenser duty varies to compensate for this).

Step 6) Determine inputs into ODE model

Once all of the equations have been put into the model, all remaining unknown variables must be placed in a section so that the user can specify these input values when running the model.

For the Excel ODE distillation model, the users inputs include:

1. Feed flow rate
2. Mole fraction of light key in the feed
3. Reflux flow rate
4. Condenser, reboiler, and tray levels
5. Phase of the feed (q-value)
6. Feed temperature
7. Integration step size

To model effects of disturbances, the user may also change these input values:

-   Feed flow after 200 time steps
-   Feed composition after 600 time steps

Step 7) Use Euler's Method to solve the ODE's

This step involves using Euler's method. to integrate each ODE over each timestep in the interval to solve for the parameter value at the next time step. Creating a plot of these values versus time allows the user to see how changes in the input values effect parameters such as distillate and bottoms composition or flowrates.

Additional Considerations for Dynamic Distillation Modeling

The Interactive Excel ODE Distillation Column Model does not take into account heat effects within or surrounding the column. It may be beneficial in modeling an actual column to determine the optimal temperatures that feed and reflux should enter at to achieve the greatest possible separation. The energy input into the reboiler is another consideration that may need to be modeled for economic purposes.

Also, to tune the dynamic distillation model with greater precision, additional parameters and equations may be added. One example of this would be to add a K_D or K_U input value to control the levels in the bottom of the column or overhead condenser.

Steady State Model

The steady state model which is described by using a McCabe-Thiele diagram shows the theoretical stages in a binary component distillation column. An example of a McCabe-Thiele diagram is shown below.

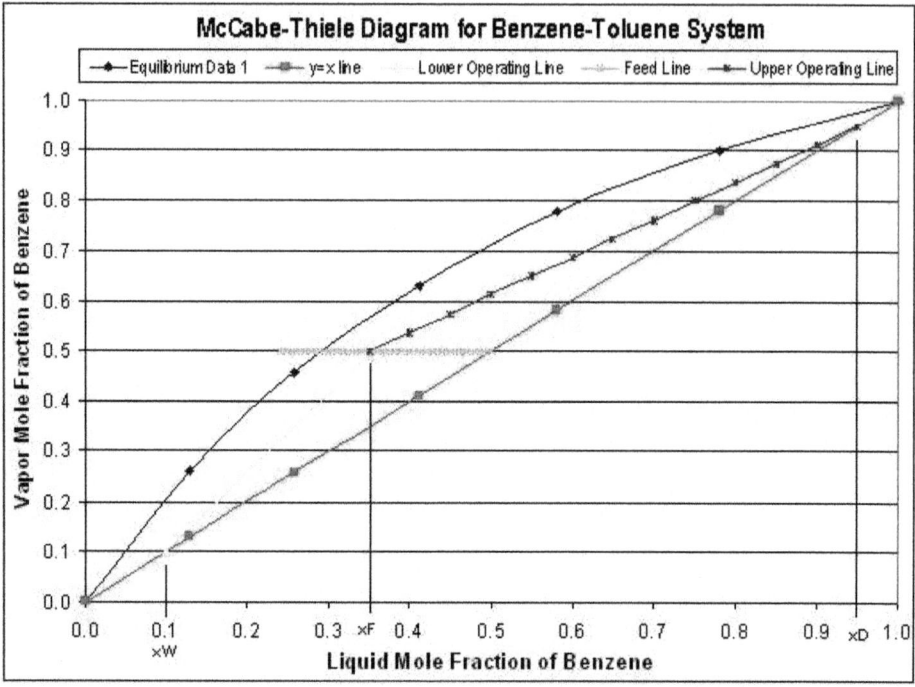

The upper operating line is a graphical representation of the vapor/liquid dynamics in each stage in the rectifying section of the column (above the feed stage), while the lower operating line represents the vapor/liquid dynamics in the stripping section of the column (below the feed stage). The starting point of the upper operating line represents the distillate composition and the bottom point of the lower operating line represents the bottoms compostion.

The feed line shows the entering feed composition as well as whether the feed is vapor, liquid, or a combination of the two. To develop this steady state model, one must know the components in the system, so that equilibrium data can be obtained. Also, this model requires that the following parameters must be known:

- reflux ratio
- distillate composition
- bottoms composition
- feed composition
- feed phase

The general equations used in the steady state model are given in the table below:

Dynamic(ODE) equations for a Simple Distillation Column

Column Location	Type of Equation	Assumption Made	Equation
All stages, but feed, condenser, &reboiler	Component M.B.	NONE	$\dfrac{dM_i x_i}{dt} = L_{i-1}x_{i-1} + V_{i+1}y_{i+1} - L_i x_i - V_i y_i$
All stages, but feed, condenser, &reboiler	Component M.B.	$\dfrac{dM_i}{dt} = 0$	$M_i \dfrac{dx_i}{dt} = L_{i-1}x_{i-1} + V_{i+1}y_{i+1} - L_i x_i - V_i y_i$
Condenser	Component M.B.	$\dfrac{dM_D}{dt} = 0$	$\dfrac{dx_{ND}}{dt} = \dfrac{1}{M_D}\left[V_R\left(y_{ND+1} - x_{ND}\right) \right]$
Rectifying Section	Component M.B.	$\dfrac{dM_i}{dt} = 0$	$\dfrac{dx_i}{dt} = \dfrac{1}{M_i}\left[L_R x_{i-1} + V_R y_{i+1} - L_R x_i - V_R y_i \right]$
Feed Tray	Component M.B.	$\dfrac{dM_{NF}}{dt} = 0$	$\dfrac{dx_{NF}}{dt} = \dfrac{1}{M_{NF}}\left[L_R x_{NF-1} + V_S y_{NF+1} + Fz_F - L_R x_{NF} - V_R y_{NF} \right]$
Stripping Section	Component M.B.	$\dfrac{dM_i}{dt} = 0$	$\dfrac{dx_i}{dt} = \dfrac{1}{M_i}\left[L_S x_{i-1} + V_S y_{i+1} - L_S x_i - V_S y_i \right]$
Re-boiler	Component M.B.	$\dfrac{dM_{NS}}{dt} = 0$	$\dfrac{dx_{NS}}{dt} = \dfrac{1}{M_B}\left[L_S x_{NS-1} - B x_{NS} - V_{reboiler} y_{NS} \right]$
Rectifying Section	Vapor Flow rates	$V_i = V_{i+1}$	$V_R = V_S + F(1-q)$
Rectifying Section	Liquid Flow rates	$L_i = L_{i+1}$	$L_R = L_D$
Stripping Section	Vapor Flow rates	$V_i = V_{i+1}$	$V_S = V_{reboiler}$
Stripping Section	Liquid Flow rates	$L_i = L_{i+1}$	$L_S = L_R + Fq$
Condenser	Total M.B.	$\dfrac{dM_D}{dt} = 0$, $V_i = V_{i+1}$ total condenser	$L_D = D + V_R$
Re-boiler	Total M.B.	$\dfrac{dM_{NS}}{dt} = 0$, $V_i = V_{i+1}$	$B = L_{NS-1} - V_{reboiler}$

All stages	Equilibrium Equation	Vapor leaving is in equilibrium w/ liquid leaving tray, Constant relative volatility	$y_i = \dfrac{\alpha x_i}{1 + (\alpha - 1)x_i}$
Re-boiler	Energy Balance	Assumes that the heat capacity of the vapor equals the heat capacity of the liquid	$\dfrac{dT_{NS}}{dt} = \dfrac{1}{M_B}(L_{NS-1}x_{NS-1} - Bx_{NS})(T_{NS-1} - T_{NS}) + \dfrac{Q_r}{M_B c_{p,l}}$
Re-boiler	Energy Balance	Used to solve for the vapor flow rate from the reboiler	$V_{reboiler} = \dfrac{1}{H_{NS-1}}(L_{NS-1}h_{NS-1} + q_{reboiler} - Wh_{NS})$
Feed Stage	Energy Balance		$\dfrac{dT_{NF}}{dt} = \dfrac{1}{M_{NF}}L_{NF-1}x_{NF-1}(T_{NF-1} - T_{NF})$ $+ V_{NF+1}y_{NF+1}(T_{NF+1} - T_{NF}) - L_{NF}x_{NF}(T_{NF-1} - T_{NF})$ $- V_{NF}y_{NF}(T_{NF+1} - T_{NF}) + Fz_F(T_{feed} - T_{NF})$
Condenser	Energy Balance	For the Excel ODE Model, the user specifies a constant reflux temperature	

Steady-state equations for a Simple Distillation Column

Column Location	Type of Equation	Equation
Condenser	Component M.B.	$f_D = (y_{ND+1} - x_{ND})$
Rectifying Section	Component M.B.	$f_i = L_R x_{i-1} + V_R y_{i+1} - L_R x_i - V_R y_i$
Feed Stage	Component M.B.	$f_{NF} = L_R x_{NF-1} + V_S y_{NF+1} + Fz_F - L_R x_{NF} - V_R y_{NF}$
Stripping Section	Component M.B.	$f_i = L_S x_{i-1} + V_S y_{i+1} - L_S x_i - V_S y_i$
Re-boiler	Component M.B.	$f_{NS} = L_S x_{NS-1} - Bx_{NS} - V_{reboiler} y_{NS}$

The McCabe-Thiele diagrams are excellent for modeling steady state operation, but they do not describe how disturbances affect the column operation. For each change in a particular parameter, a separate McCabe-Thiele diagram must be made. The dynamic model, although more complex than the steady state model, shows how a column operates during start-ups, when disturbances occurs, and where steady state conditions occur. Therefore, the dynamic ODE model of a distillation column can allow the user to how product purities and flow rates change with time.

Glossary of Terms

M_i = Molar holdup on tray i

L_{i-1} = Liquid molar flowrate into tray i

L_i = Liquid molar flowrate leaving tray i

V_{i+1} = Vapor molar flowrate entering tray i

V_i = Vapor molar flowrate leaving tray i

x_i = mole fraction of light component in the Liquid phase of Tray i

y_i = mole fraction of light component in the Gas phase of Tray i

B = Bottoms flowrate

D = Distillate flowrate

f = Feed flow rate

α = Relative volatility of Benzene-Toluene system.

q = Vapor Liquid compostion value

OUR EXCEL DISTILLATION MODEL

Below you will find a link for our distillation model in Excel. It is an exceptionally large download so it may take some time if you are not using a high speed internet connection. Because it takes some time for the iterations to run to completion, calculations are not continuous within the spreadsheet. Be patient as it may take a few seconds for the calculations to run when operational parameters change. Also, because of the nature of the iterative process Excel uses to determine the cell values, large deviations may cause the model to crash resulting in #Error# for many cells. If this occurs, close the model and re-open it from the website to try different parameters.

In order to demonstrate dynamic changes occuring during operation, step changes are incorporated into the Excel model. To show the effects of a feed change, input the change into cell C-18. This change will occur at t = 2min for a length 2 minutes and then the feed flowrate will revert back to the original value. To demonstrate a change in the feed compostion at t = 6min, input the change into cell C-19 on the Excel model. This compostion change will be constant over the remainder of the time and will not revert to the previous value. Notice the changes occuring in the graphs as these step changes occur.

This model has a specified control range because typical operation of distillation columns requires operation within a specific range of values where input parameters marginally vary. It is recommended to use the following variables for the initial model.

Feed Flow to Column = 50.00

Feed Composition (xF) = 0.5

Reflux Flow = 15

Bottoms Flow @ T= 0 = 35.00

Mtray = 10

Mcondenser = 50

Mreboiler = 50

qF = 0.4

Feed Temperature = 80

Reflux Temp = 80

Initial Column Temp = 80

Qreboiler = 100

Avoid extreme changes in the initial values, otherwise a significant error may occur causing the model to crash. The principal cause of this is due to the heat effects which are taken into account in the model and the fact that the number of trays on the column is set at 6.

Terms Commonly Used in Distillation

Active tray area: The region of the tray where the upward moving vapor comes into contact with the downward flowing liquid (where the mass transfer occurs)

Downcomer: Area on side of trays where liquid flows down through the distillation column

Disturbance (wrt distillation): Any minor change in the distillation column caused by an external or internal source that causes product variability

Flooding: Liquid from the active tray area is carried up into the vapor stream (occurs at low L/V ratios)

Ratio Control: Controlled ratio of two manipulated variables

Tray Fouling: Active area of tray is deminished, thus reducing separation efficiency within the column

Upset: Any major change in the distillation caused by an external source that produces erratic column operation that requires manual override to gain control

Weeping Points: Liquid from the active tray area seeps downward through the tray instead of flowing through the downcomer(occurs at high L/V ratios)

HEAT EXCHANGE MODEL

In process industries, heat exchangers are designed to transfer heat from one fluid to another. Heat exchangers have many different applications, especially in chemical processes, air conditioning, and refrigeration. They are classified by their design and type of flow. In some types of heat exchangers, the two fluids are separated by a wall or membrane and the heat transfer takes place by both convection and conduction. In another, less common type of exchanger, the two fluids physically come into contact with each other as the heat transfer occurs.

Since heat exchangers have a wide variety of applications and are commonly used in industry, control of the system is essential. A dynamic model may be created to allow the chemical engineer to optimize and control the heat exchanger. By utilizing this model, predictions can be made about how altering the independent variables of the system will change the outputs. There are many independent variables and considerations to account for in the model. If done so correctly, accurate predictions can be made about the system.

Types of Heat Exchangers

1. Double - Pipe Heat Exchanger

A double-pipe heat exchanger is the simplest type of heat exchanger and can operate with co-current or counter-current flow. The design consists of a single small pipe (tube-side) inside of a larger one (shell-side). A co-current heat exchanger is most commonly used when you want the exiting streams to leave the exchanger at the same temperature. A counter-current heat exchanger is used more often than co-current because they allow for a more efficient transfer of energy.

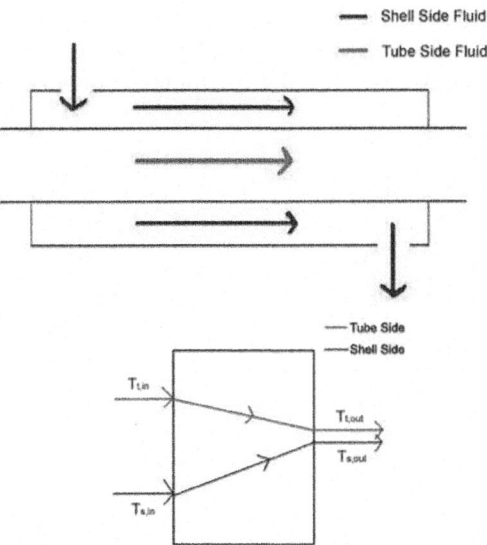

Fig. : Co-current Flow.

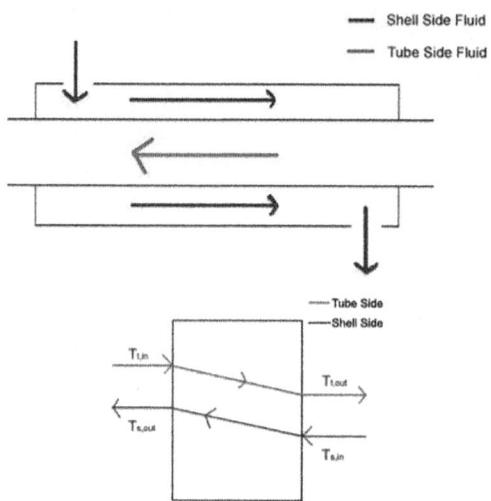

Fig. : Counter current Flow.

2. Shell-and-Tube Exchanger

A shell-and-tube exchanger is used for larger flows, which are very common in chemical process industries. The design of this exchanger is a shell with a bundle of tubes inside. The tubes are in parallel and a fluid flows around them in the shell. There are many different arrangements such as straight or u-tube. Each arrangement allows for a different type of flow such as co-current, counter-current and cross flow.

The tube-side can have one or more passes to increase the energy exchange from the tube-side fluid. The shell-side may contain baffles, or walls, that channel the fluid flow and induce turbulence, and thus, increase energy exchange. Correlations can be developed to predict the increase in energy exchange.

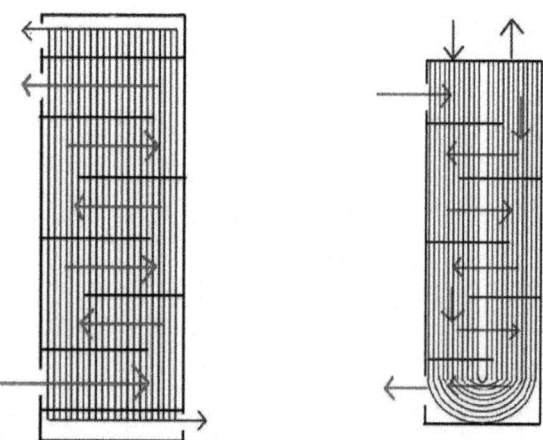

Fig. : Straight Tube Heat Exchanger.   **Fig.** : U-Tube Heat Exchanger.

3. Cross-flow Exchanger

The most common application for a cross-flow heat exchanger is when a gas is being heated or cooled. This device consists of multiple tubes in parallel, usually containing a liquid, and the exterior gas flows across the tubes. In some cases the air is confined in separate flow chambers, such as fins, and in others it is open to flow freely.

Adiabatic PFR

A plug flow reactor (PFR) is a tubular reactor used in chemical reactions. Reactants enter one end of the PFR, while products exit from the other end of the tube. A PFR is useful because of high volumetric conversion and good heat transfer. PFRs carry out power law reactions faster than CSTRs and generally require less volume. Excel modeling for an adiabatic plug flow reactor is useful for estimating conversion as a function of volume. By inputting the values of the constants and iteratively changing the volume, the conversion at a specific volume can be determined.

This removes the need for complicated hand calculations and helps the user visualize the reaction and notice trends through the PFR. Alternatively, excel modeling can also determine the temperature inside the reactor as a function of volume. Euler's integration is used to estimate the conversion or temperature through the PFR. By ensuring that the volume intervals are small, the conversion or temperature values generated should be fairly accurate. Refer to the Euler's method section for more information on its implementation.

Plug Flow Reactor

F_{A0}
C_{A0}
T_0

F_A
F_B
C_A
C_B
T

Basic Algorithm to Model Adiabatic PFR

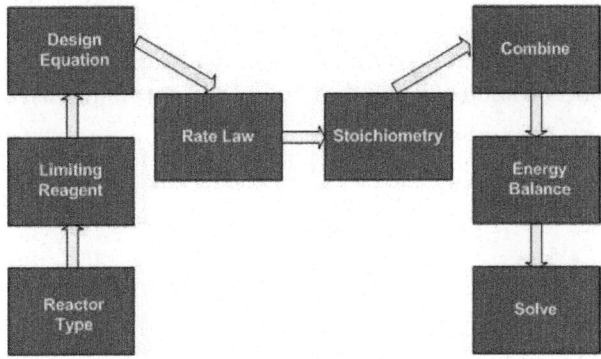

The model presented will include the following assumptions:

- Negligible pressure drop
- Single elementary reaction
- Power law kinetics
- Constant fluid properties (*i.e.* heat capacity)

To describe the basic algorithm for an adiabatic PFR, a simple case will be considered where pure A enters the reactor:

$$aA< - - >bB$$

1. Reactor Type

PFR

2. Limiting Reagent

The design and rate equations should be expressed in terms of the limiting reagent. Here, A is the limiting reagent.

3. Design Equation

The design equation for a PFR can be expressed in terms of several variables, including conversion, moles, and concentration. It is important to note that the design equation in terms of conversion may only be used if one reaction is occuring. Otherwise, it must be expressed in terms of moles or concentration. The equation to describe conversion as a function of volume, as derived from a simple mole balance, is shown below in Equations.

$$\frac{dX}{dV} = \frac{-r_A}{F_{Ao}}$$

X= conversion

V= volume

r_A = reaction rate of A

F_{Ao} = initial moles of A

4. Rate Law

The rate law is independent of reactor type. It is expressed in terms of a rate constant and concentration. If the reaction is irreversible, the rate law is modeled according to Equation. If the reaction is reversible, the rate law models Equation 3 in which the concentration of B and the equilibrium constant must be accounted for.

$$-r_A = kc_A^a$$

$$-r_A = k\left(c_A^a - \frac{c_B^b}{K_c}\right)$$

k= rate constant

c_A^a = concentration of A with stoichiometric coefficient *a*

c_B^b = concentration of B with stoichiometric coefficient b

K_c = equilibrium constant

If the reaction is not isothermal, the rate constant and equilibrium constant should be written in terms of temperature as shown in Equation 4 and 5.

$$k = k_0 exp(\frac{E}{R}(\frac{1}{T_0} - \frac{1}{T}))$$

$$K_C = K_{C1} exp(\frac{\Delta H_{RX}}{R}(\frac{1}{T_1} - \frac{1}{T}))$$

k_0 = rate constant at T_0

E = activation energy

R = ideal gas constant

T = temperature

K_{C1} = equilbrium constant at T_1

ΔH_{RX} = standard heat of reaction

5. Stoichiometry

Stoichiometry is used to express concentration of a species in terms of concentration of the limiting reactant. The stoichiometry equation depends on whether the reaction occurs in the liquid or gas phase. If it occurs in the gas phase, Equations are used. For liquid phase, ε becomes zero and the correlation reduces to Equation. There is no pressure term included because it is assumed that pressure drop is negligible, and the initial pressure is equal to the final pressure.

$$C_A = C_{Ao} * \frac{(1 - X)}{(1 + \epsilon X)} \frac{T_0}{T}$$

$$C_B = C_{Ao} * \frac{(\theta_B + \frac{b}{a}X)}{(1 + \epsilon X)} \frac{T_0}{T}$$

$$C_A = C_{Ao} * (1 - X)$$

$$C_B = C_{Ao} * (\theta_B + \frac{b}{a}X)$$

ε = y_{Ao} * (stoichiometric coefficients of products - coefficients of reactants)

y_{Ao} = initial mole fraction of A

θ_i = ratio of initial moles (or concentration) of species i to initial moles (or concentration) of A

6. Combine

The design equation, rate law, and stoichiometric equations are combined. For example, assume the reaction is irreversible, liquid phase, and isothermal, with pure A entering the reactor and has the combined equation as shown below.

$$\frac{dX}{dV} = \frac{k * C_{A0}^a (1 - X)^a}{F_{A0}}$$

7. Energy Balance

The energy balance can be used to relate temperature and conversion. It assumes heat flow and shaft work are not present.

$$T = \frac{X(-\Delta H_{RX}) + \Sigma \Theta_i C_{pi} T_o + X \Delta C_p T_r}{\Sigma \Theta_i C_{pi} + X \Delta C_p}$$

C_{pi}= heat capacity of species i

ΔCp= Cp products- Cp reactants

If ΔCp= 0, the energy balances reduces to the following:

$$T = \frac{X(-\Delta H_{RX}) + \Sigma \Theta_i C_{pi} T_o}{\Sigma \Theta_i C_{pi}}$$

8. Solve

Solve the system of equations in Excel. The ODE is approximated by Euler's method.

Using the Excel Model

On the "Reaction Scheme" worksheet enter the various stoichiometric values for your reaction, the specific heats for the various reactants and products, and also the heat of reaction and the reference temperature that the heat of reaction is taken at. The "Input Conditions" worksheet is fairly straightforward. If the reaction is gas-phase make sure to enter a "1" in the appropriate cell and if liquid-phase enter a "0".

The "Rate Law" sheet contains the definitions for the values you will have to enter on this sheet. Note that T1 is the reference temperature for your rate constant. Make sure to enter a "1" in the cell if your reaction is reversible, and a "0" if it is irreversible. Make sure not to tamper with the "Numerical Calculations" worksheet! You can, however, scroll downward to extract exact values for volume, conversion, and temperature down the PFR.

Multiple reactions are not included in this model. Having multiple reactions would require a seperate reaction scheme sheet for every reaction, and would require many different adjustments to ensure that all scenarios are accounted for.

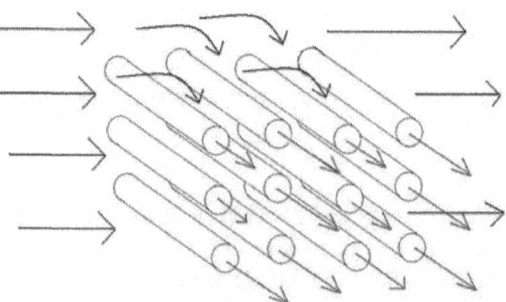

Fig. : Cross-Flow Heat Exchanger (free flow).

Dynamic Modeling of Heat Exchangers

Since heat exchangers are so widely used in industry, it necessary for a chemical engineer to be able to optimize and control the system and know how independent variables will affect the outputs from the system. To do this, a dynamic model is developed and utilized.

A dynamic model of a heat exchanger may be used, for example, to predict how a change in the fluid flowrates or the addition of an insulating jacket will affect the outlet temperature of the product stream. The model uses ordinary differential equations (ODEs) to describe the process and, using a program like Microsoft EXCEL, gives plots of the variables vs. time for the entire process. There are many independent variables in a heat exchanger, which can cause modeling to be very complex since multiple ODEs are required to define all of the process variables.

Some of the independent variables in a heat exchanger system include:

- Shell-side:
  - o fluid
  - o flowrate
  - o temperature
  - o number of baffles
- Tube-side:
  - o fluid
  - o flowrate
  - o temperature
  - o number of passes
- Flow configuration
  - o Co-current
  - o Counter-current
  - o Cross flow
- Insulating jacket

The primary dependent variable of concern is the outlet temperature of the product stream (usually the tube-side fluid. The outlet temperature, in a controlled system, is monitored by a sensor. The sensor then transmits a signal to an actuating device of one or more of the independent variables (usually shell-side flow controller) to perform some desired response.

Modeling with ODEs

The following section outlines the method for developing a dynamic model of a heat exchanger. The model is for a double-pipe heat exchanger that has the ability to flow in co-current or counter-current configurations and the option of an external insulating jacket. It assumes constant fluid properties and perfect heat transfer through the metal of the tubing. The outlet temperature of the tube-side fluid is monitored by a temperature sensor, and the flow rate of the shell-side fluid is controlled by a actuated flow-controlling device.

Energy Balance

An energy balance is first performed on the tube-side fluid.

$$\begin{array}{c} \text{Rate of} \\ \text{accumulation} \\ \text{of thermal} \\ \text{energy in} \\ \text{tube-side} \\ \text{fluid} \end{array} = \begin{array}{c} \text{Rate of} \\ \text{energy} \\ \text{in} \end{array} - \begin{array}{c} \text{Rate of} \\ \text{energy} \\ \text{out} \end{array} - \begin{array}{c} \text{Heat} \\ \text{transferred} \\ \text{from} \\ \text{shell-side} \end{array}$$

The left-most term in the above energy balance is the amount of thermal energy that accumulates in the tube-side fluid and causes a change in its outlet temperature. The terms on the right side of the above energy balance describe the thermal energy of the fluid flowing in and the fluid flowing out and the amount of heat transfer from the shell side fluid. In the term for heat transfer from the shell-side, the temperatures are the temperatures of the outlet streams. The outlet temperatures will change according to whether you are running co-currently or counter-currently. The energy balance is written as:

$$mc_{p,t}\frac{dT_{t,out}}{dt} = \rho c_{p,t}F_{t,in}T_{t,in} - \rho c_{p,t}F_{t,out}T_{t,out} - \frac{kA_i}{\Delta z}(T_{t,out} - T_{s,out})$$

where,

m = mass of the fluid = $V\rho = \rho A_{crosssectional}\Delta z$

c_p = constant pressure heat capacity of the fluid

T = Temperature

t = time

k = conductive heat transfer coefficent

A = surface area of tube that fluid contacts

Δz = length of tube

ρ = density of the fluid

F = volumetric flowrate of the fluid

and subscripts denote

t - tube-side fluid

out - outlet

in - inlet

i - inside

An similar energy balance is next performed on the shell-side fluid.

Rate of accumulation of thermal energy in shell-side fluid	=	Rate of energy in	-	Rate of energy out	-	Rate of heat transferred to tube-side fluid	-	Rate of heat loss to the surroundings

The left-most term in the above energy balance is the amount of thermal energy that accumulates in the shell-side fluid and causes a change in its outlet temperature. The terms on the right side of the above energy balance describe the thermal energy of the fluid flowing in and the fluid flowing out, heat transfer to the tube-side fluid, and also the heat lost by convection to the surroundings. The energy is written as;

$$mc_{p,s}\frac{dT_{s,out}}{dt} = \rho c_{p,s}F_{s,in}T_{s,in} - \rho c_{p,s}F_{s,out}T_{s,out} - \frac{kA_o}{\Delta z}(T_{s,out} - T_{t,out}) - hA_{o'}(T_s - T_\infty)$$

where,

h = coefficient of convective heat transfer for air

k - conductive heat transfer coefficient.

and subscripts denote

s - shell-side fluid

∞ - air

o - outside of tube

o' - outside of shell

Considerations

There are considerations and simplifications you can make in order to solve the differential energy balances. The validity of these assumptions depends on how accurate of a model you require.

1. The heat capacity of the fluid may be temperature dependent. If this is the case, a polynomial equation could be written for the C_p values of each of the fluids. It would take the form-

$$C_p = a + bT + cT^2 + dT^3$$

Values of a, b, c, and d are properties of the fluid and can be found in *Perry's Chemical Engineers' Handbook*.

It should also be noted that if the fluids in the process are gases, their fluid properties will be affected by changes in pressure and the use of a constant C_p value would be inappropriate.

2. The density of the fluid may be temperature dependent. This would be likely if the fluid were a vapor, as in the case of using steam as the shell-side fluid to heat the tube-side process fluid. If this is the case, a differential equation could be written for the ρ value of the fluid and would take the form-

$$\frac{d\rho}{dT} \propto C$$

where;

C - is the coefficient of cubic expansion (relates kinetic energy to temperature).

3. Heat loss to the surroundings may be neglected. This would be the case if the heat exchanger is well insulated or if the shell-side fluid is about the same as ambient temperature. In this case-

$$hA(T_s - T_\infty) = 0$$

4. The temperature sensor may have an inherent time delay. This means the temperature output from the sensor lags the actual temperature at the time of the reading. The temperature lag may be accounted for with the differential equation-

$$\frac{dT_{outlet,sensor}}{dt} = \frac{1}{\tau_{Ts}}(T_{outlet,actual} - T_{outlet,sensor})$$

where,

τ_{Ts} = time constant for the temperature sensor

τ_{Ts} is a process parameter and usually ranges from 6 to 20 seconds depending on the properties of the sensor. This value would either be given in literature from the manufacturer or would have to be determined through experimentation.

5. The actuator system for the control valve may have a slow dynamic response. When the actuator system is based on a control valve, the response to the pressure change is slower than the flow through the valve. The flow rate through a control valve can be modeled with the differential equation-

$$\frac{dQ_{actual}}{dt} = \frac{1}{\tau_v}(Q_{setpoint} - Q_{actual})$$

where,

τ_v = time constant for the flow control valve

τ_v is a process parameter and usually ranges from 0.5 to 2 seconds depending on the properties of the sensor. This value would have to be determined through experimentation.

Using Excel to Solve ODEs

Since temperature of the heat exchanger varies across the metal as well as along the length of the pipe, a partial derivative for how temperature varies with length must be described. In order to solve this problem using excel, an approximation of length was used. Instead of taking the partial derivative across the length with respect to temperature, the pipe was divided into differential segments, Δz. Ideally Δz is an infinitesimally small cross section of the length of the heat exchanger.

We will assume that through this differential segment, the temperature of the liquid that leaves the segment is the same as the temperature of the liquid within the segment. Since we are assuming the same temperature for the exiting streams as the inside of the segment, the choice of length for these Δz's helps dictate the accuracy of the solution. The combination of these differential units allows us to model a heat exchanger without the use of partial derivatives. The figure below shows an example of the simplification where the heat exchanger is split into three segments.

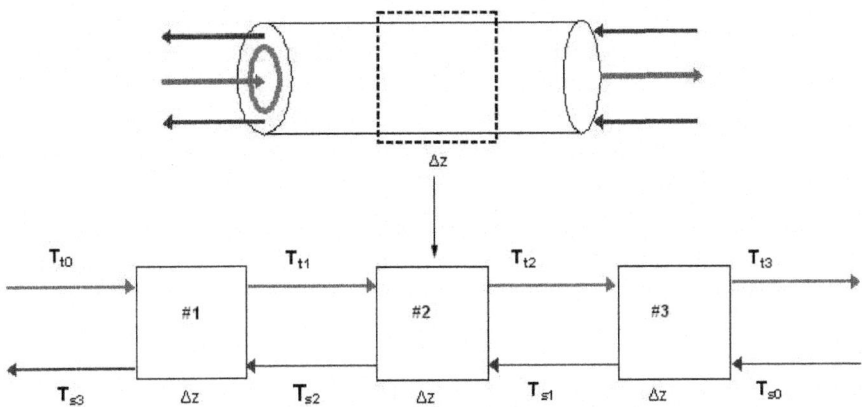

Fig. : Heat Exchanger Simplification.

To begin modeling, start by solving the appropriate energy balance for $\dfrac{dT}{dt}$ and make any simplifications necessary. Then use a method to solve ODEs in Excel, such as Euler's or Heun's.

For each unit, Δz, and each fluid in that unit of the heat exhanger, you should have an approximation equation. Because ideally Δz represents an infinitesimal section, it follows from the first assumption that we can take the temperature in Δz as being the exit temperature of the hot and cold streams for the Δz's respectively.

The equation for Euler's Method takes the form:

$$T_{h1}(t_{i+h}) = T_{h1}(t_i) + \Delta t\left[\frac{dT}{dt}(t_i)\right]$$

The equation for Heun's Method takes the form:

$$T_{h_1}(t_{i+h}) = T_{h_1}(t_i) + \Delta t\left[\frac{1}{2}\frac{dT}{dt}(t_i, T_i) \quad + \quad \frac{1}{2}\frac{dT}{dt}(t_{i+h}, T_{i+hk_i})\right]$$

The temperatures can then be plotted versus time in order to model how the system inputs affect heat exchange.

Electric Vehicle Cruise Control

Controls principles developed in this course can be applied to non-chemical engineering systems such as automobiles. Some companies, such as NAVTEQ, are developing adaptive cruise control products that use information about the upcoming terrain to shift gears in a more intelligent manner which improves speed regulation and fuel economy. This case study will examine the basics of develop a speed controller for an electric vehicle.

An electric vehicle was chosen for the following reasons:

- Electric vehicles are interesting from an engineering perspective and may become a reality for consumers in the future

- Torque produced by an electric motor is instantaneous (for all practical purposes). Thus actuator lag can be ignored, simplifying the development of said controller.

- Some electric vehicles feature motors directly integrated into the hub of the drive wheel(s). This eliminates the need for a transmission and simplifies vehicle dynamics models.

Forces

As shown in the free body diagram below, there are six forces acting on the vehicle:

1. Rolling Resistance
2. Aerodynamic Drag
3. Aerodynamic Lift
4. Gravity
5. Normal
6. Motor

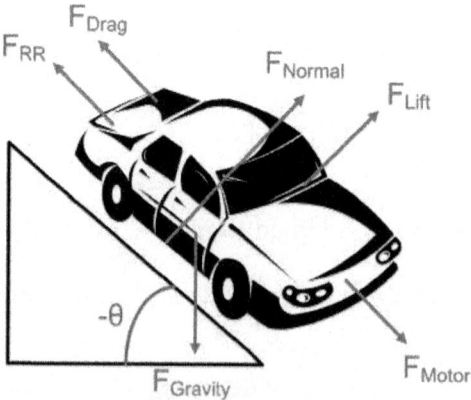

Fig. : A free body diagram of the forces acting on the vehicle. $-\theta$ is used to denote the grade of the road such that a positive value of θ corresponds to the vehicle traveling uphill.

Rolling Resistance

Rolling resistance is due the tires deforming when contacting the surface of a road and varies depending on the surface being driven on. It can be model using the following equation:

$$F_{RR} = C_{rr1}v + C_{rr2}F_N$$

The two rolling resistance constants can be determined experimentally and may be provided by the tire manufacture. F_N is the normal force.

Aerodynamic Drag

Aerodynamic drag is caused by the momentum loss of air particles as they flow over the hood of the vehicle. The aerodynamic drag of a vehicle can be modeled using the following equation:

$$F_{drag} = (1/2)\,\rho CdAv^2$$

- ρ is the density of air. At 20°C and 101kPa, the density of air is 1.2041 kg/m^3.
- CdA is the coefficient of drag for the vehicle times the reference area.
- v is the velocity of the vehicle.

Aerodynamic Lift

Aerodynamic lift is caused by pressure difference between the roof and underside of the vehicle. Lift can be modeled using the following equation:

$$F_{lift} = (1/2)\,\rho ClAv^2$$

- ρ is the density of air. At 20°C and 101kPa, the density of air is 1.2041 kg/m^3.
- ClA is the coefficient of lift for the vehicle times the reference area.
- v is the velocity of the vehicle.

Gravity

In the diagram above, there is a component of gravity both in the dimension normal to the road and in the dimension the vehicle is traveling. Using simple trigonometry, the component in the dimension of travel can be calculated as follows:

$$F_{G,travel} = mg\,sin(-\theta)$$

- m is the mass of the vehicle.
- g is the acceleration due to gravity.

Normal Force

The normal force is the force excerted by the road on the vehicle's tires. Because the vehicle is not moving up or down (relative to the road), the magnitude of the normal forces equals the magnitude of the force due to gravity in the direction normal to the road.

$$F_N = F_{G,norm.} - F_{lift} = mg \cos(-\theta) - (1/2)\rho ClAv^2$$

- m is the mass of the vehicle.
- g is the acceleration due to gravity.

Motor

The torque produced by an electric motor is roughly proportional to the current flowing through the stater of the motor. In this case study, the current applied to the motor will be controlled to regulate speed. Applying a negative current will cause the vehicle to regeneratively brake.

$$\tau = k_{motor}\, I$$
$$F_M = \frac{\tau}{r} = \frac{k_{motor}I}{r}$$

- τ is the torque produced by the motor.
- I is the current flowing through the motor.
- r is the radius of the tire.
- k_{motor} is a constant.

Newton's Second Law

Using Newton's Second Law, a differential for the vehicle's speed can be obtained.

$$ma = \sum F = F_M - F_{drag} - F_{RR} + F_{G,travel}$$

Substituting in the expressions for various forces detailed above yields the following:

$$ma = \frac{\tau}{r} - (1/2)\rho CdAv^2 - C_{rr1}v - C_{rr2}F_N + mg\sin(-\theta)$$

Further substituting the expression for normal forces yields the following:

$$ma = \frac{\tau}{r} - (1/2)\rho CdAv^2 - C_{rr1}v - C_{rr2}[mg\cos(-\theta) - (1/2)\rho ClAv^2] + mg\sin(-\theta)$$

Substituting $\frac{dv}{dt}$ for a results in the following:

$$m\frac{dv}{dt} = \frac{\tau}{r} - (1/2)\rho CdAv^2 - C_{rr1}v - C_{rr2}[mg\cos(-\theta) - (1/2)\rho ClAv^2] + mg\sin(-\theta)$$

Grouping like turns results in the following:

$$m\frac{dv}{dt} = \frac{\tau}{r} - [(1/2)\rho CdA - C_{rr2}(1/2)\rho ClA]v^2 - C_{rr1}v - C_{rr2}mg\cos(-\theta) + mg\sin(-\theta)$$

In order to simply the remaining analysis, several constants are defined as follows:

$$\alpha = (1/2)\,\rho r(CdA - A_{rr2}ClA)$$

$$\beta = rC_{rr1}$$

$$\gamma = rC_{rr2}mg\cos(-\theta) = rmg\sin(-\theta)$$

Substituting these into the differential equation results in the following expression:

$$\frac{dv}{dt} = \frac{\tau - \alpha v^2 - \beta v - \gamma(\theta)}{mr}$$

It is important that θ (and thus also γ) is a function of vehicle position.

PID Controller

One way to regulate vehicle speed is to control the torque generated by the electric motor. If the motor torque is greater than the resistive torque acting on the vehicle (a summation of aerodynamic drag, rolling resistance, *etc.* the vehicle will accelerate. If the motor torque is less than the resistive torque, the vehicle will slow down.

Expression for Phase Current

For an electric motor, the phase current flowing through the motor is proportional to the torque produced. Thus one strategy for controlling the vehicles speed is to controller the motor phase current.

Using a PID controller architecture, the expression for motor current is the following:

$$I = K_c(v_{set} - v) + \frac{1}{\tau_I}\int (v - vset)dt + \tau_D \frac{d(v_{set} - v)}{dt} + C_{offset}$$

Differential Equation for Velocity

Substituting this expression into the differential equation for vehicle position results in the following:

$$\frac{dv}{dt} = \frac{k_{motor}\left[K_c(v_{set} - v) + \frac{1}{\tau_I}\int (v_{set} - v)dt + \tau_D \frac{d(v_{set} - v)}{dt} + C_{offset}\right] - \alpha v^2 - \beta v - \gamma(\theta)}{mr}$$

Defining another variable x_1 allows for the removal of the integral from the expression.

$$\frac{dx_1}{dt} = v_{set} - v$$

$$\frac{dv}{dt} = \frac{k_{motor}\left[K_c(v_{set} - v) + \frac{x_1}{\tau_I} + \tau_D \frac{d(v_{set} - v)}{dt} + C_{offset}\right] - \alpha v^2 - \beta v - \gamma(\theta)}{mr}$$

If all changes in v_{set} are gradual then $\frac{dv_{set}}{dt} \approx 0$. Applying this simplification results in the following expression:

$$\frac{dv}{dt} = \frac{k_{motor}[K_c(v_{set} - v) + \frac{x_1}{\tau_I} - \tau_D \frac{dv}{dt} + C_{offset}] - \alpha v^2 - \beta v - \gamma(\theta)}{mr}$$

Using a little algebra, an expression for $\frac{dv}{dt}$ can be obtained as follows:

$$\frac{dv}{dt} = \frac{k_{motor} K_c(v_{set} - v) + \frac{k_{motor}}{\tau_I} x_1 + k_{motor} C_{offset} - \alpha v^2 - \beta v - \gamma(\theta)}{k_{motor} t_D + mr}$$

Find Fixed Point

The fixed point can be obtained by setting the derivatives to zero and solving the system of equation.

System of Equations:

$$\frac{dv}{dt} = \frac{k_{motor} K_c(v_{set} - v) + \frac{k_{motor}}{\tau_I} x_1 + k_{motor} C_{offset} - \alpha v^2 - \beta v - \gamma(\theta)}{k_{motor} t_D + mr} = 0$$

$$\frac{x_1}{dt} = v_{set} - v = 0$$

Solution:

$$x_1 = \frac{-\tau_I (k_{motor} C_{offset} - \alpha v_{set}^2 - \beta v_{set} - \gamma(\theta)])}{k_{motor}}$$

$$v = v_{set}$$

As expected, a fixed point exists when the set velocity equals the actual velocity of the vehicle.

Linearize System of ODEs

Before the stability of the system can be (easily) examined, the system must be linearized around a fixed point.

Note: To simply the matrix expressions in section the following notation will be used:

$$y_i' = \frac{dy_i}{dt}$$

Overall a linearized system of ODEs has the following form:

$$\begin{bmatrix} y_1' \\ y_2' \\ \vdots \\ y_n' \end{bmatrix} = J \begin{bmatrix} y_1 \\ y_2 \\ \vdots \\ y_n \end{bmatrix} + \begin{bmatrix} k_1 \\ k_2 \\ \vdots \\ k_n \end{bmatrix}$$

The first step of linearizing any system of ODEs to calculate the Jacobian. For this particular system, the Jacobian can be calculated as follows:

$$J = \begin{bmatrix} \frac{\partial v'}{\partial v} & \frac{\partial v'}{\partial x_1} \\ \frac{\partial x_1'}{\partial v} & \frac{\partial x_1'}{\partial x_1} \end{bmatrix} = \begin{bmatrix} \frac{-k_{motor}K_c - mr[2\alpha v + \beta]}{mr + k_{motor}\tau_D} & \frac{k_{motor}}{t_I(mr + k_{motor}\tau_D)} \\ -1 & 0 \end{bmatrix}$$

The Jacobian is then evaluated at the fixed point:

$$J = \begin{bmatrix} \frac{-k_{motor}K_c - mr[2\alpha v_{set} + \beta]}{mr + k_{motor}\tau_D} & \frac{k_{motor}}{t_I(mr + k_{motor}\tau_D)} \\ -1 & 0 \end{bmatrix}$$

The next step if to calculate the vector of constants. For this particular system, said vector can be calculated as follows:

$$\begin{bmatrix} k_v \\ k_{x_1} \end{bmatrix} = -J \begin{bmatrix} v \\ x_1 \end{bmatrix} \Bigg|_{fixedpoint} = -\begin{bmatrix} \frac{-k_{motor}K_c - 2\alpha v_{set} - \beta}{mr + k_{motor}\tau_D} & \frac{k_{motor}}{t_I(mr + k_{motor}\tau_D)} \\ -1 & 0 \end{bmatrix} \begin{bmatrix} v_{set} \\ \frac{-\tau_I(k_{motor}C_{offset} - \alpha v_{set}^2 - \beta v_{set} - \gamma(\theta)])}{k_{motor}} \end{bmatrix}$$

$$\begin{bmatrix} k_v \\ k_{x_1} \end{bmatrix} = \begin{bmatrix} \frac{-(-k_{motor}K_c - 2\alpha v_{set} - \beta])v_{set}}{(mr + k_{motor}\tau_D)} + \frac{(k_{motor}C_{offset} - \alpha v_{set}^2 - \beta v_{set} - \gamma(\theta)])}{(mr + k_{motor}\tau_D)} \\ v_{set} \end{bmatrix}$$

Combining the Jacobian and vector of constants results in the following linearized system:

$$\begin{bmatrix} v' \\ x_1' \end{bmatrix} = J \begin{bmatrix} v \\ x_1 \end{bmatrix} + \begin{bmatrix} k_v \\ k_{x_1} \end{bmatrix}$$

$$\begin{bmatrix} v' \\ x_1' \end{bmatrix} = \begin{bmatrix} \frac{-k_{motor}K_c - 2\alpha v_{set} - \beta}{m + k_{motor}\tau_D} & \frac{k_{motor}}{t_I(mr + k_{motor}\tau_D)} \\ -1 & 0 \end{bmatrix} \begin{bmatrix} v \\ x_1 \end{bmatrix} + \begin{bmatrix} \frac{-(-k_{motor}K_c - 2\alpha v_{set} - \beta])v_{set}}{(mr + k_{motor}\tau_D)} + \frac{(k_{motor}C_{offset} - \alpha v_{set}^2 - \beta v_{set} - \gamma(\theta)])}{(mr + k_{motor}\tau_D)} \\ v_{set} \end{bmatrix}$$

Stability Analysis

To assess the stability of the controller, the eigenvalues of the Jacobian in the linearized systems of ODEs can be examined. In general, an eigenvalue (*lambda*) is the solution to the following equation:

$$|J - \lambda I| = 0$$

Using a computer to solve said equation, the eigenvalues of this particular system can be found to be the following:

$$\lambda = \frac{k_{motor}K_c\tau_I + \beta\tau_I + 2\alpha v_{set}\tau_I \pm \sqrt{\tau_I(k_{motor}^2 K_c^2\tau_I + 2k_{motor}K_c\tau_I\beta + 4k_{motor}K_c\tau_I\alpha v_{set} + \beta^2\tau_I + 4\beta\tau_I\alpha v_{set} + 4\alpha^2 v_{set}^2\tau_I - 4k_{motor}mr - 4k_{motor}^2 t_D)}}{2\tau_I(mr + k_{motor}\tau_D)}$$

For the system to be stable, the real component of all eigenvalues must be non-positive. The following inequality must be true for a stable controller:

$$k_{motor}K_c\tau_I + \beta\tau_I + 2\alpha v_{set}\tau_I < -Real(\tau_I(k_{motor}^2 K_c^2\tau_I + 2k_{motor}K_c\tau_I\beta + 4k_{motor}K_c\tau_I\alpha v_{set} + \beta^2\tau_I + 4\beta\tau_I\alpha v_{set} + 4\alpha^2 v_{set}^2\tau_I - 4k_{motor}mr - 4k_{motor}^2 t_D))$$

For the system to not oscillate, the imaginary component of all eigenvalues must be zero. The following inequality must be true for a non-oscillating controller:

$$0 > \tau_I(k_{motor}^2 K_c^2\tau_I + 2k_{motor}K_c\tau_I\beta + 4k_{motor}K_c\tau_I\alpha v_{set} + \beta^2\tau_I + 4\beta\tau_I\alpha v_{set} + 4\alpha^2 v_{set}^2\tau_I - 4k_{motor}mr - 4k_{motor}^2 t_D)$$

Interestingly, neither of these criteria depend on the grade of the road (θ). However, during the analysis, it was assumed that θ is constant. For most roads, this is not the case; θ is actually a function of vehicle position. In order to add this additional level of detail, the original system of ODEs needs to be revised:

$$\frac{dv}{dt} = \frac{k_{motor}K_c(v_{set} - v) + k_{motor}x_1 + k_{motor}C_{offset} - \alpha v^2 - \beta v - \gamma(\theta)}{k_{motor}t_D + mr}$$

$$\frac{dx_1}{dt} = v_{set} - v$$

$$\frac{ds}{dt} = v$$

$$\theta = f(s)$$

Unfortunately for any normal road, the grade is not a simple (or even explicate) function of position (s). This prevents an in depth analytical analysis of stability. However for a very smooth road with very gradual changes in grade, the stability of the controller should be unaffected.

Example Electric Vehicle

Simulating the system also for other properties of the controller to be examined. This section presents the results from simulating a specified fictitious electric vehicle.

Parameters

For the fictitious electric vehicle simulated in this analysis, the following parameters were used. These parameters are roughly based on the parameters one would expect to see in a typical electric automobile.

$$m = 1400 \ kg$$
$$CdA = 0.58 \ m^2$$
$$ClA = 0 \ m^2$$
$$C_{rr1} = 2.75\frac{Ns}{m}0.018$$
$$C_{rr2} = 0.018$$
$$k_{motor} = 1.2\frac{Nm}{A_{rms}}$$

Root Locus Plots

Using the parameters above, Root Locus Plots of the system were constructed to numerically explore the stability. The following controller constants were used when constructing the plots:

$$vset = 25 \ m/s$$
$$K_c = 55$$
$$\tau_I = 0.5$$
$$\tau_D = 1$$
$$C_{offset} = 0$$

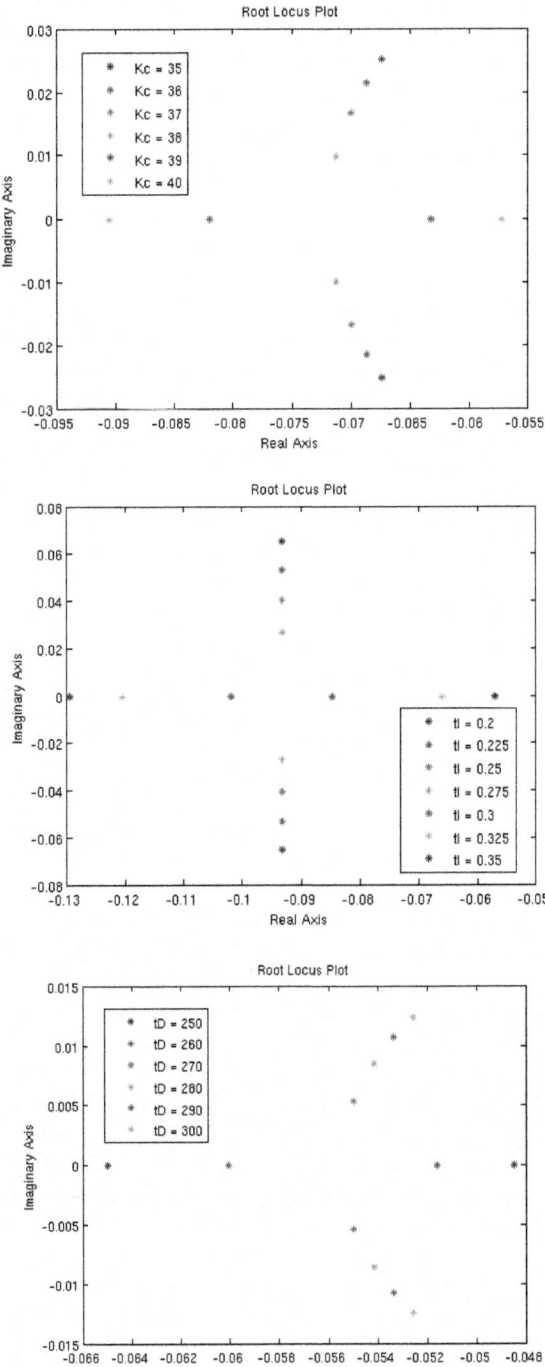

These Root Locus plots show that the controller with said constants is both stable and does not oscillate.

Phase Portrait

A phase portrait of the system with the following parameters was also constructed:

$$v_{set} = 25 \, m/s$$
$$K_c = 55$$
$$\tau_I = 0.5$$
$$\tau_D = 20$$
$$C_{offset} = 0$$

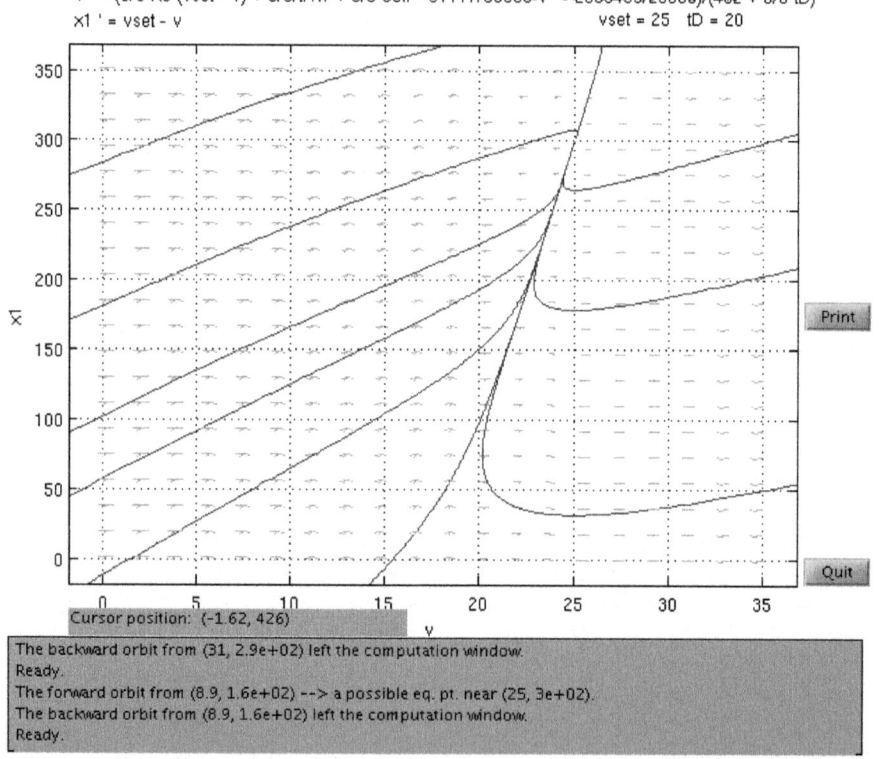

The phase portrait shows that the system is both stable and does not oscillate, as predicted by the Root Locus plots.

Driving Simulation on Level Terrain

The vehicle was simulated starting at 10 m/s and accelerating to 25 m/s via cruise control on level terrain ($\theta = 0$). For this simulation, the following constants were used:

$$v_{set} = 25 \, m/s$$

$$K_c = 55$$

$$\tau_I = 0.5$$

$$\tau_D = 20$$

$$C_{offset} = 146 \approx \frac{\alpha v_{set}^2 + \beta v_{set} + \gamma(\theta = 0)}{k_{motor}}$$

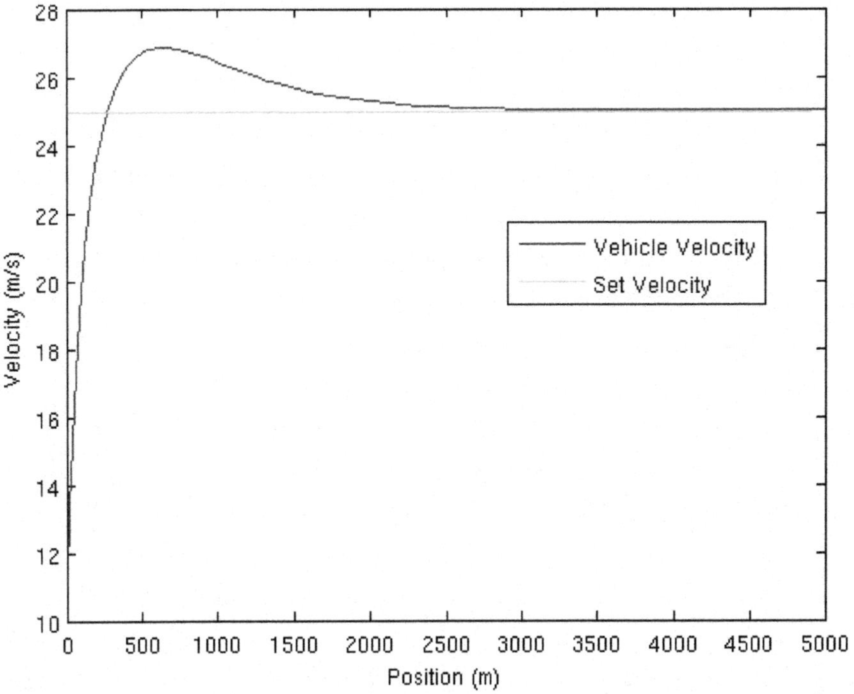

Interestingly this graph shows oscillation, despite the Root Locus plots and phase diagrams. It is important to remember, however, that the Root Locus plot and stability methods involve linearizing the system. It is possible the linearized system is not a good approximation of the system.

It is also important to remember that this particular example involves a large set point change, which can induce oscillations in certain systems.

Driving Simulation on Unlevel Terrain

In order to explore how the controller behaves on a road with a non-zero grade, a route with hills was constructed from the following equation, where h is elevation and s is position.:

$$h = 100 \sin\left(\frac{s}{250\pi}\right) + \frac{s}{2000}$$

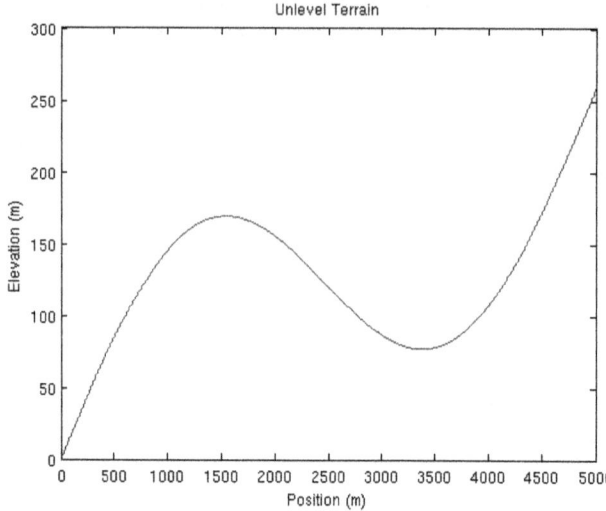

The vehicle was simulated driving said road starting with the following initial conditions and controller constants:

Initial Conditions

$$s(t=0) = 0$$
$$x_1(t=0) = 0$$
$$v(t=0) = v_{set} = 25 \, m/s$$

Controller Constants

$$k_c = 55$$
$$\tau_I = 0.5$$
$$\tau_D = 20$$
$$C_{offset} = 146$$

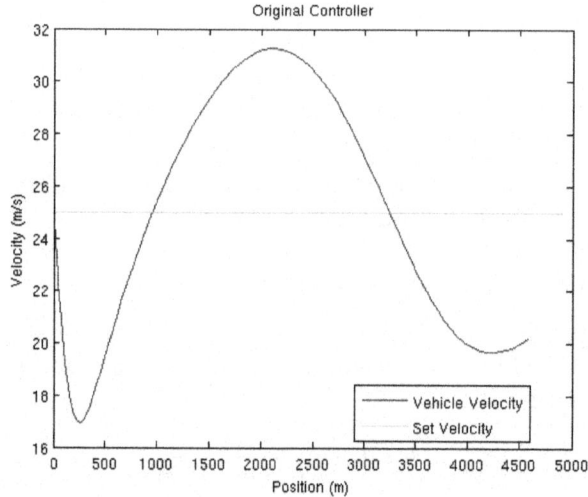

The current controller tuning is inadequate for this road. There are vary large variations in velocity. In order to reduce these variation, the propartional gain K_c was increased by a factor of 5. Below are the results:

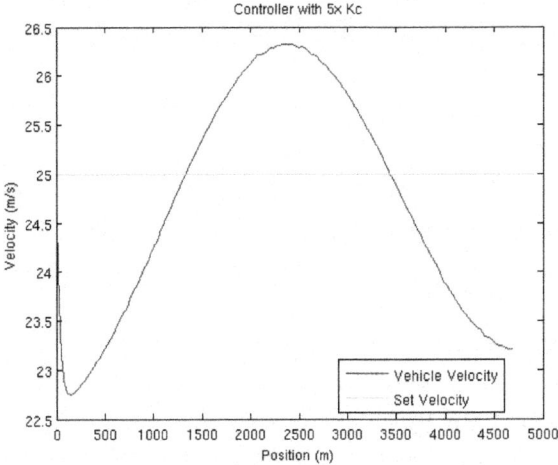

Using the Optimization Toolbox in MATLAB, the controller was optimized by minimizing the sum of the errors (difference between vehicle velocity and set velocity) on this particular segment of road. Below are the optimized controller constants and a plot of the velocity profile:

$$k_c = 232.58$$
$$\tau_I = 0.001$$
$$\tau_D = 0$$
$$C_{offset} = 114.62$$

This example shows the power of optimization and model predictive control.

Biology Application

Background of why Insulin Control is Important for Diabetic Patients

Diabetes mellitus is a disease of the endocrine system where the body cannot control blood glucose levels. There are two general classifications of diabetes:

Type I (also known as juvenile diabetes)

- Genetic predisposition and/or an autoimmune attack destroys T-cells of pancreas
- Body cannot produce insulin to regulate blood glucose

Type II

- Most common form of diabetes and has reached epidemic status in the United States
- Usually caused by lifestyle
- Obesity reduces body's responsiveness to insulin

Treatment for both types of diabetes may include exercise, dieting, oral medications, or insulin injections. Most insulin dependent diabetics follow a management plan that requires frequent testing of blood glucose levels and then injection of a prescribed dose of insulin based on the blood glucose level. However, the downside of this treatment method is that there is no predictive control. If blood glucose levels are falling and insulin is administered, a hypoglycemic episode may occur.

Recent biomedical advancements have resulted in continuous blood glucose monitoring devices as well as insulin pumps. Continuous monitoring allows for finer blood glucose control and can help predict fluctuations in the blood glucose level.

Insulin pumps replace the need to administer insulin injections by automatically injecting a prescribed dose, however it requires blood glucose level input from the patient. In the future, insulin pumps and continuous blood glucose monitors may be integrated forming a closed loop control system which will can replace the body's own faulty control system.

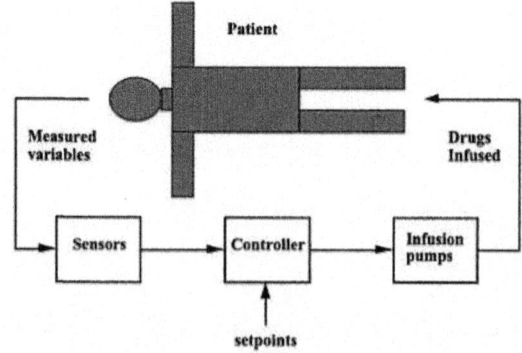

Fig. : Control Schematic of Insulin Infusion.

Mathematical Model for a Closed Loop Insulin Delivery System

The following set of differential equations are known as the Bergman "minimal model":

$$\frac{dG}{dt} = -p_1 G - X(G - G_b) + \frac{G_{meal}}{V_1}$$

$$\frac{dX}{dt} = -p_2 X + p_3 I$$

$$\frac{dI}{dt} = -n(I + I_b) + \frac{U}{V_1}$$

Where:

G = deviation variable for blood glucose concentration

X = deviation variable for insulin concentration in a "remote" compartment

I = deviation variable for blood insulin concentration

G_{meal} = a meal disturbance input in glucose

U = the manipulation insulin infusion rate

G_b = steady state value of blood glucose concentration

I_b = steady state value of blood insulin concentration

Blood parameters include p_1, p_2, p_3, n, V_1(blood volume). These are specific to the blood specimen and must be predetermined.

A linear state space model can be used to express the Bergman equations seen above. The general form for a state space model can be seen below:

$$\begin{bmatrix} \dot{x}_1 \\ \vdots \\ \dot{x}_n \end{bmatrix} = \begin{bmatrix} a_{11} & \cdots & a_{1n} \\ \vdots & \ddots & \vdots \\ a_{n1} & \cdots & a_{nn} \end{bmatrix} \begin{bmatrix} x_1 \\ \vdots \\ x_n \end{bmatrix} + \begin{bmatrix} b_{11} & \cdots & b_{1m} \\ \vdots & \ddots & \vdots \\ b_{n1} & \cdots & b_{nm} \end{bmatrix} \begin{bmatrix} u_1 \\ \vdots \\ u_m \end{bmatrix}$$

$$\begin{bmatrix} y_1 \\ \vdots \\ y_r \end{bmatrix} = \begin{bmatrix} c_{11} & \cdots & c_{1n} \\ \vdots & \ddots & \vdots \\ c_{r1} & \cdots & c_{rn} \end{bmatrix} \begin{bmatrix} x_1 \\ \vdots \\ x_n \end{bmatrix} + \begin{bmatrix} d_{11} & \cdots & d_{1m} \\ \vdots & \ddots & \vdots \\ d_{r1} & \cdots & d_{rm} \end{bmatrix} \begin{bmatrix} u_1 \\ \vdots \\ u_m \end{bmatrix}$$

In general:

$\dot{x} = Ax + Bu$

$y = Cx + Du$

Where:

x = states

u = inputs

y = outputs

$$A = \begin{bmatrix} -p_1 & -G_b & 0 \\ 0 & -P_2 & P_3 \\ 0 & 0 & -n \end{bmatrix}$$

$$B = \begin{bmatrix} 0 & \frac{1}{V_1} \\ 0 & 0 \\ \frac{1}{V_1} & 0 \end{bmatrix}$$

$$C = \begin{bmatrix} 1 & 0 & 0 \end{bmatrix}$$

$$D = \begin{bmatrix} 0 \\ 0 \end{bmatrix}$$

Using this general formula, we can deconstruct the Bergman equations as a linear state space model. The first input is the insulin infusion and the second input represents the meal glucose disturbance.

First Input:

$$\begin{bmatrix} \dot{G} \\ \dot{X} \\ \dot{I} \end{bmatrix} = \begin{bmatrix} -p_1 & -G_b & 0 \\ 0 & -P_2 & P_3 \\ 0 & 0 & -n \end{bmatrix} \begin{bmatrix} G \\ X \\ I \end{bmatrix} + \begin{bmatrix} 0 & \frac{1}{V_1} \\ 0 & 0 \\ \frac{1}{V_1} & 0 \end{bmatrix} \begin{bmatrix} u_1 \\ u_2 \end{bmatrix}$$

Where

$\dot{G}$ = differential blood glucose concentration

$\dot{X}$ = differential insulin concentration in a "remote" compartment

$\dot{I}$ = differential blood insulin concentration

Second Input:

$$y = \begin{bmatrix} 1 & 0 & 0 \end{bmatrix} G + \begin{bmatrix} 0 & 0 \end{bmatrix} \begin{bmatrix} u_1 \\ u_2 \end{bmatrix}$$

Where:

$$\begin{bmatrix} u_1 \\ u_2 \end{bmatrix} = \begin{bmatrix} U - U_b \\ G_{meal} - 0 \end{bmatrix}$$

Example Set of Parameters

Gb = 4.5mmol/liter

Ib = 4.5mU/liter

V1 = 12 liters

p1 = 0/min

p2 = 0.025/min

p3 = 0.0000013mU/liter

n = 5/54 min-1

It is important to keep track of units when using these parameters. The concentrations listed above are in mmol/liter, but the glucose disturbance has units of grams. Therefore, it is necessary for us to apply a conversion factor of 5.5556mmol/grams to the G_{meal} term. Using these variables solve one can solve for the steady states, and calculate the basal insulin infusion rate (Ub) such that is is equal to 16.1667mU/min. For these parameters, the resulting state space model is

$$A = \begin{bmatrix} 0 & -4.5 & 0 \\ 0 & -0.025 & 0.000013 \\ 0 & 0 & \frac{-5}{54} \end{bmatrix}$$

$$B = \begin{bmatrix} 0 & 0.4630 \\ 0 & 0 \\ \frac{1}{12} & 0 \end{bmatrix}$$

It is common practice in the U.S. to describe glucose concentration in units of mg/deciliter as opposed to mmol/liter. Therefor, the units will be converted from mmol/liter to mg/deciliter. The molecular weight of glucose is 180g/mol, and therefore one it is necessary to multiply the glucose state (mmol/liter) by 18, so that the measured glucose output obtained will be in units of mg/deciliter. The following state-output relationship will handle that:

$$C = \begin{bmatrix} 18 & 0 & 0 \end{bmatrix}$$

$$D = \begin{bmatrix} 0 & 0 \end{bmatrix}$$

Through Laplace transforms it will be found that the process transfer function is :

$$G_p(s) = \frac{-3.79}{(40s + 1)(10.8s + 1)s}$$

The disturbance transfer function due to pole/zero cancellation is simply :

$$G_d(s) = \frac{8.334}{s}$$

However, in reality glucose does not directly enter the blood stream. There is a "lag time" associated with the processing of glucose in the gut. It must first be processed here before entering the blood. However, it can be modeled as a first-order function, with a 20-minute time constant. This modifies the above equation for the disturbance transfer function to include the lag in the gut such that:

$$G_d(s) = \frac{8.334}{s(20s + 1)}$$

Desired Control Performance

The steady-state glucose concentationof 4.5mmol/liter corresponds to the glucose concentration of 81 mg/deciliter. A diabetic patient, in order to stay healthy, must keep their blood glucose concentration above 70 mg\deciliter. If the blood glucose concentration falls below 70 mg/deciliter the patient may be likely to faint, as this is a very typical short-term symptom of hypoglycemia.

The insulin infusion rate (maniuplated input) cannot fall below 0, and this constraint must be set in simulations. Numerous different control strategies can be used. The equation given for the process function above can be simplified down to the following form and a PD or PID controller can be used.

$$G_p(s) = \frac{k_p}{(tau_p + 1)s}$$

A diabetic will know when they are consuming a meal, and therefore when their blood glucose concentration may rise. Therefore a feed-forward control system may be desired.

Chapter 6

PIPING AND INSTRUMENTATION DIAGRAMS

PID GENERAL INFORMATION

Piping and Instrumentation Diagrams (P&IDs) use specific symbols to show the connectivity of equipment, sensors, and valves in a control system. This chapter will outline general information about P&IDs that is necessary to know before trying to draw one.

P&ID *vs.* PFD

P&IDs may often be confused with PFDs, or process flow diagram. P&IDs and PFDs generally utilize the same notation for equipment. However, they serve different purposes and provide different information. The purpose of a PFD is to show exactly what a process does during operation, and a P&ID shows all controllers, valve types and the materials that are used in construction. A PFD shows the connectivity and relationships between the major equipment and materials in a process.

However, it also includes tabulated design values such as normal, minimum, and maximum operating conditions that a P&ID does not contain. A PFD does not include minor piping systems or other minor components that a P&ID typically includes. The difference between P&IDs and PFDs is that P&IDs typically include more information regarding piping and safety relief valves than process flow diagrams. P&IDs do not contain operating specifications that PFDs contain, such as stream flows and compositions.

It is important to note that differences between PFDs and P&IDs will vary between institutions. Most corporations maintain designated standards to create and modify the documents. Both PFDs and P&IDs are controlled documents and need to be maintained with a document control procedure.

PFD/PID INDUSTRY EXAMPLE

Process Flow Diagrams *vs.* Process and Instrumentation Diagrams

Process Flow Diagrams PFD	Process and Instrumentation Diagram P&ID
• General process flows between major equipment clearly presented in a simplified manner	• General process flows present but convoluted due to many sheets and abundance of information
• Omits valves, controls and minor pipelines	• Includes all valves, controls and minor pipelines
• Provides baseline for operating conditions for flows, compositions, etc. (min, normal, max)	• Information on piping (material of construction and schedule) and safety features (i.e. pressure relief valves, interlocks)
• Stream labeled with information presented at the bottom of each sheet	• Safe operating ranges of equipment presented at bottom of each sheet

Process Description

This PFD/P&ID is taken from a brewery which is planned to produce 100,000 bottles a day of beer. This section includes boiling and fermentation. The process starts by using steam to boil the beer in KB-301, which sterilzes the beer. Next, the beer is pumped into a whirlpool filter where impurities are removed. The beer is then cooled to 15 degrees C, using a heat exchanger E-306. Once the beer is cooled, it is sent to the fermentation tank, TK-307. Here yeast is added, which metablozies sugar in the beer into alcohol and carbon dioxide. After a couple days, the beer is sent to a maturation tank, TK-314. The beer spends about 4 days in this tank, before it is clarified and bottled.

Process Flow Diagrams

A PFD for the process is here:

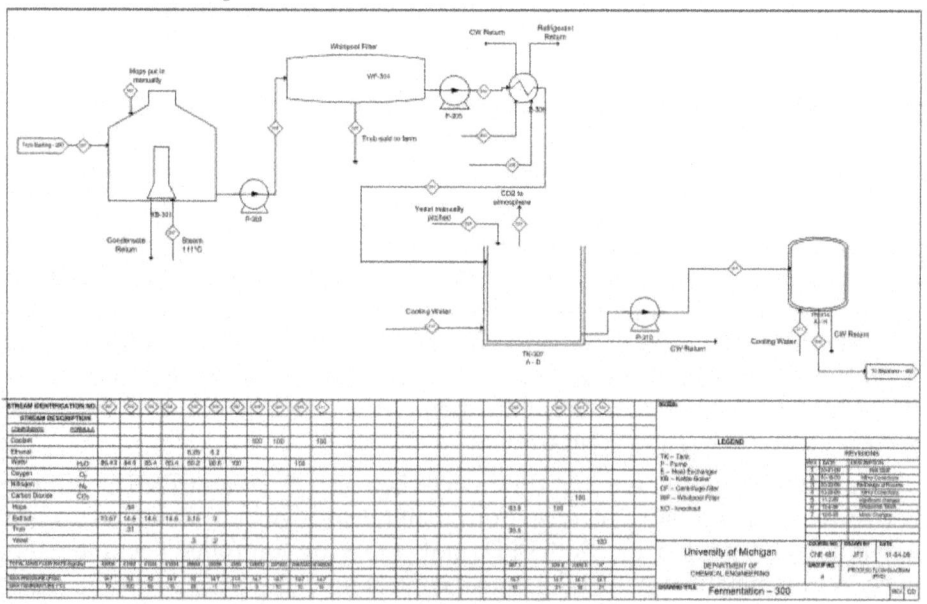

Image courtesy of John Trumble, University of Michigan

The PFD contains stream compositions on the border, instead of the equipment. The overall flow rates through each stream can be seen on the PFD also.

Process Flow Diagrams

A P&ID for the process is here:

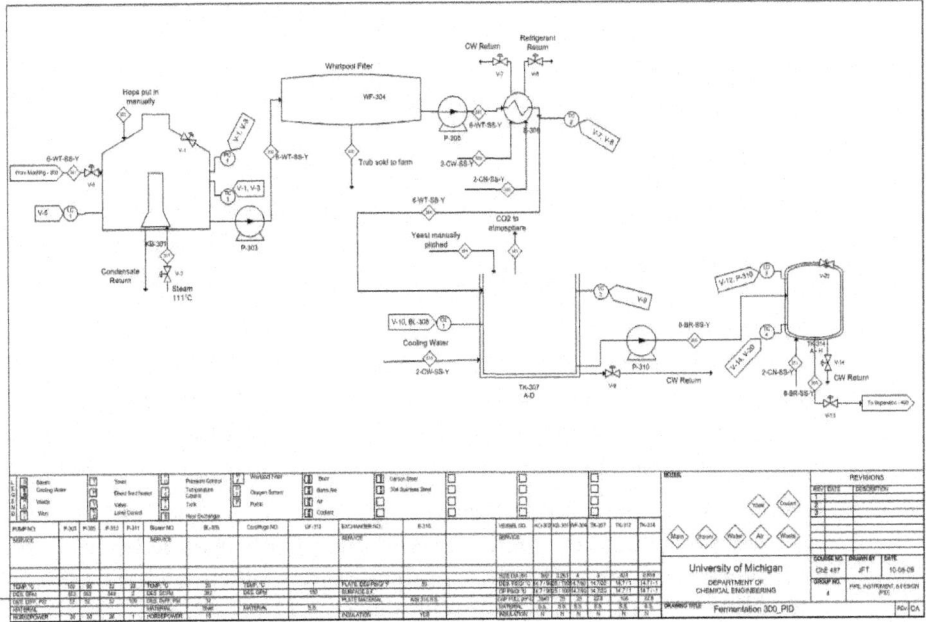

Information Incorporated in P&IDs

The following information is given on a P&ID that is not explicit on a PFD:

- ALL valves and valve types
- Controllers present
- Controller architectures
- Pipe diameters, materials of construction, and insulation properties (including minor piping systems)
- Equipment materials of construction

Uses of P&IDs

- Develop operational methodology
- Develop safety philosophy and safeguards
- Develop control philosophy
- Serve as a basis for control programming

- Serve as a communication document for how the process works
- Serve as a basis for equipment design, pipe design, estimating cost, purchasing
- Use for evaluation of construction process
- Train employees
- Serve as a conceptual layout of a chemical plant
- Provide a common language for discussing plant operations

Characteristics of P&IDs

- Grouped by specific section of plant
- Show the connections between all the sensors and actuators
- A general schematic - NOT a layout and NOT to scale. It should be noted that P&IDs do not specifically imply the following: same elevation of equipment, relative sizes, where valves are located on the equipment, how close equipment is to each other, and impeller types/location. They are also not the same as control or incidence diagrams. This type of information can be seen in either a plant layout drawing (can be either satellite view, showing distance between units, or a slice of building, showing height of units) or construction drawings, such as plant blueprints.
- Must be clear and uncluttered
- Must be systematic and uniform. P&IDs are used extensively in industry to document plant information. These documents need to be easily read by anyone working within the company, and easily explained to anyone else. OSHA audits can occur anytime and it is imperative that operational information can be provided to the auditor when requested. Without standard notation, it would be very difficult to go from plant to plant within your company and understand the P&IDs.
- Are generally highly confidential and have restricted access, as they can be used to replicate a process

What A P&ID Is Not

- Not an architectural diagram of a process (should show the flow of material across the plant floor between sensors and actuators, not necessarily corresponding to a 3D location)
- Does not need to be drawn perfectly to scale
- Does not imply any relative elevations
- Do not need manual switches
- No actual temperature, pressure, or flow data
- Leave out any extensive explanations

What A P&ID Should Include

- Instrumentation and designations
- Mechanical equipment with names and numbers, and their specifications such as material, insulation, maximum flow rate, working pressure and temperature, maximum power *etc.*
- All valves and their identifications
- Process piping, sizes and identification
- Miscellaneous - vents, drains, special fittings, sampling lines, reducers, increasers and swagers
- Permanent start-up and flush lines
- Flow directions
- Interconnections references
- Control inputs and outputs
- Interfaces for class changes
- Vendor and contractor interfaces
- Identification of components and subsystems delivered by others
- Intended physical sequence of the equipment

P&ID Revisions

- Revisions should be clearly identified
- Regularly issued to all related employees at each significant change to the process, as well as at benchmark points
- If small changes are made that don't warrant a completely new revision, "red pencil" additions are generally accepted between issues
- 15-20 revisions are typical during process design
- All revisions need to be communicated to EVERYONE so that only the latest revision is used. This is critical in order to avoid serious (not to mention expensive) construction mistakes. (Typically outdated P&ID is discarded to avoid confusion)

PID STANDARD NOTATION

Piping and Instrumentation Diagrams (P&IDs) use specific symbols to show the connectivity of equipment, sensors, and valves in a control system. These symbols can represent actuators, sensors, and controllers and may be apparent in most, if not all, system diagrams. P&IDs provide more detail than a process flow diagram with the exception of the parameters, *i.e.* temperature, pressure, and flow values.

"Process equipment, valves, instruments and pipe lines are tagged with unique identification codes, set up according to their size, material fluid contents, method of connection (screwed, flanged, *etc.* and the status (Valves - Normally Closed, Normally Open)."These two diagrams can be used to connect the parameters with the control system to develop a complete working process. The standard notation, varying from letters to figures, is important for engineers to understand because it a common language used for discussing plants in the industrial world.

P&IDs can be created by hand or computer. Common programs, for both PC and Mac, that create P&IDs include Microsoft Visio (PC) and OmniGraffle(Mac). As with other P&IDs, these programs do not show the actual size and position of the equipment, sensors and valves, but rather provide a relative positions. These programs are beneficial to produce clean and neat P&IDs that can be stored and viewed electronically.

This chapter covers four main types of nomenclature. The first section describes the use of lines to describe process connectivity. The second section describes letters used to identify control devices in a process. The third section describes actuators, which are devices that directly control the process. The final section describes the sensors/transmitters that measure parameters in a system.

Line Symbols

Line symbols are used to describe connectivity between different units in a controlled system. The table describes the most common lines.

Table: Line Symbols

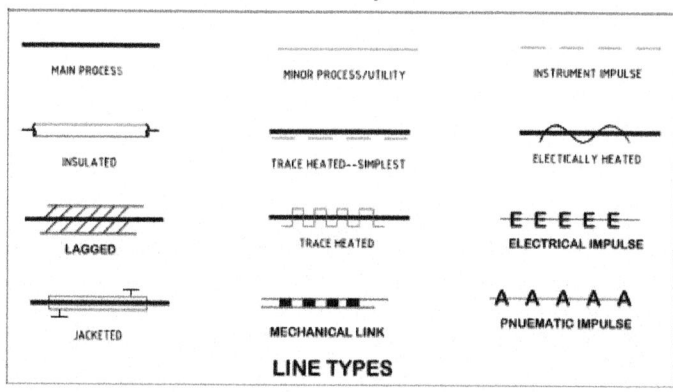

In Table, the "main process" refers to a pipe carrying a chemical. "Insulated" is straightforward, showing that the pipe has insulation. "Trace heated" shows that the pipe has wiring wrapped around it to keep the contents heated. The last column in Table shows pipes that are controlled by a controller. "Electrical impulse" shows that the manner in which information is sent from the controller to the pipe is by an electrical signal, whereas "pneumatic impulse" indicates information sent by a gas.

In addition to line symbols, there are also line labels that are short codes that convey further properties of that line. These short codes consist of: diameter of pipe, service, material, and insulation. The diameter of the pipe is presented in inches. The service is what is being carried in the pipe, and is usually the major component in the stream. The material tells you what thatsection of pipe is made out of. Examples are CS for carbon steel or SS for stainless steel. Finally a 'Y' designates a line with insulation and an 'N' designates one without it. Examples of line short codes on a P&ID are found below in Figure.

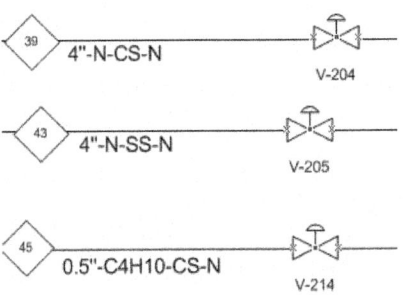

Fig. A: Line Labels.

This is useful for providing you more practical information on a given pipe segment.

For example in stream 39 in Figure, the pipe has a 4" diameter, services/ carries the chemical denoted 'N', is made of carbon steel, and has no insulation.

Identification Letters

The following letters are used to describe the control devices involved in a process. Each device is labeled with two letters. The first letter describes the parameter the device is intended to control. The second letter describes the type of control device.

Table : First Identification Letter.

First Letter	Parameter Controlled
A	Analysis
C	Conductivity
D	Density
E	Voltage
F	Flow rate
I	Current
L	Level
M	Moisture (Humidity)
P	Pressure or Vacuum
T	Temperature
V	Viscosity

Second Letter	Type of Control Device
A	Alarm
C	Control
I	Indicate
T	Transmit
V	Valve

For example, the symbol "PI," is a "pressure indicator."

Valve Symbols

The following symbols are used to represent valves and valve actuators in a chemical engineering process. Actuators are the mechanisms that activate process control equipment.

Table: Valve Symbols.

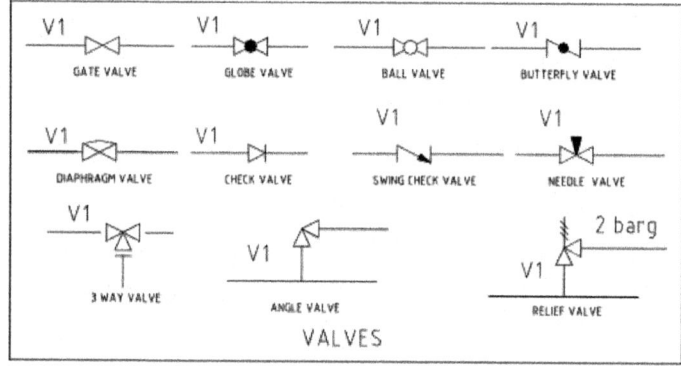

Table: Valve Actuator Symbols.

General Instrument or Function Symbols

Instruments can have various locations, accessibilities, and functionalities in the field for certain processes. It is important to describe this clearly in a P&ID. Below is a table of these symbols commonly used in P&IDs.

General instrument or function symbols			
	Primary location accessible to operator	Field mounted	Auxiliary location accessible to operator
Discrete instruments	1	2	3
Shared display, shared control	4	5	6
Computer function	7	8	9
Programmable logic control	10	11	12

1. Symbol size may vary according to the user's needs and the type of document.
2. Abbreviations of the user's choice may be used when necessary to specify location.
3. Inaccessible (behind the panel) devices may be depicted using the same symbol but with a dashed horizontal bar.

Source: Control Engineering with data from ISA S5.1 standard

Discrete instruments are instruments separate or detached from other instruments in a process. Shared display, shared control instruments share functions with other instruments. Instruments that are controlled by computers are under the "computer function" category. Instruments that compute, relay, or convert information from data gathered from other instruments are under the "Programmable logic control" section.

For example, a discrete instrument for a certain process measures the flow through a pipe. The discrete instrument, a flow transmitter, transmits the flow to a shared display shared control instrument that indicates the flow to the operator. A computer function instrument would tell the valve to close or open depending on the flow. An instrument under the "Programmable logic control" category would control the valve in the field if it was pneumatically controlled, for instance. The instrument would gather information from discrete instruments measuring the position of the actuator on the valve, and would then adjust the valve accordingly.

In the chart above, it is necessary to know where the instrument is located and its function in order to draw it correctly on a P&ID. A primary instrument is an instrument that functions by itself and doesn't depend on another instrument. A field mounted instrument is an instrument that is physically in the field, or the plant. Field mounted instruments are not accessible to an operator in a control room. An auxiliary instrument is an instrument that aids another primary or auxiliary instrument. Primary and auxiliary instruments are accessible to operators in a control room.

Transmitter Symbols

Transmitters play an important role in P&IDs by allowing the control objectives to be accomplished in a process. The following are commonly used symbols to represent transmitters.

Below are three examples of flow transmitters. The first is using an orifice meter, the second is using a turbine meter, and the third is using an undefined type of meter.

Table: Transmitter Symbols.

Flow Transmitters		
FT	FT	FT
Orifice Meter	Turbine Meter	Undefined Meter

The location of the transmitter depends on the application. The level transmitter in a storage tank is a good example. For instance, if a company is interested in when a tank is full, it would be important for the level transmitter to be placed at the top of the tank rather than the middle. If the transmitter was misplaced in the middle because a P&ID was misinterpreted then the tank would not be properly filled. If it is necessary for the transmitter to be in a specific location, then it will be clearly labeled.

Miscellaneous Symbols

The following symbols are used to represent other miscellaneous pieces of process and piping equipment.

Table: Process Equipment.

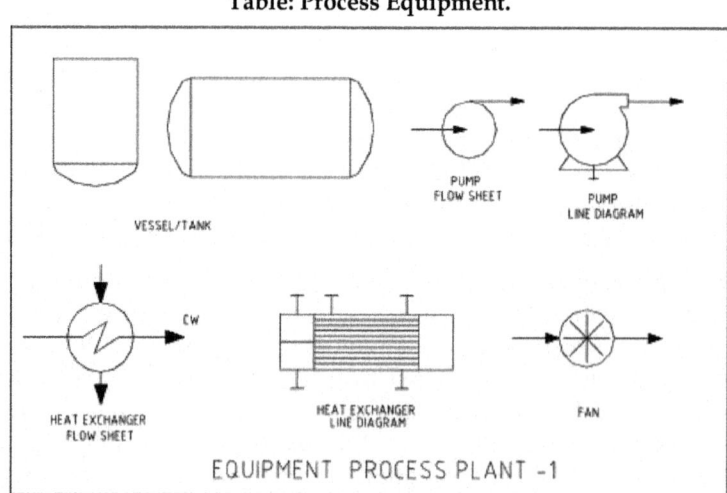

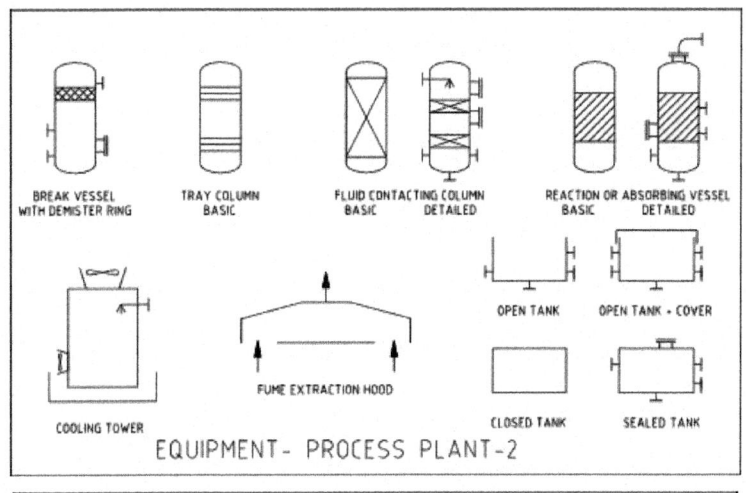

EQUIPMENT- PROCESS PLANT-2

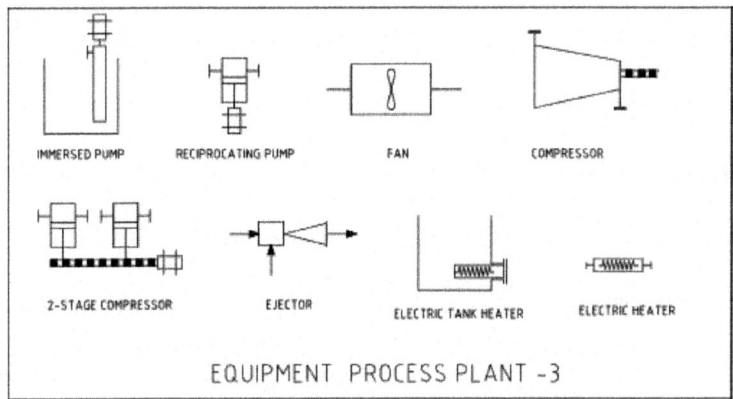

EQUIPMENT PROCESS PLANT -3

Table: Line Fittings.

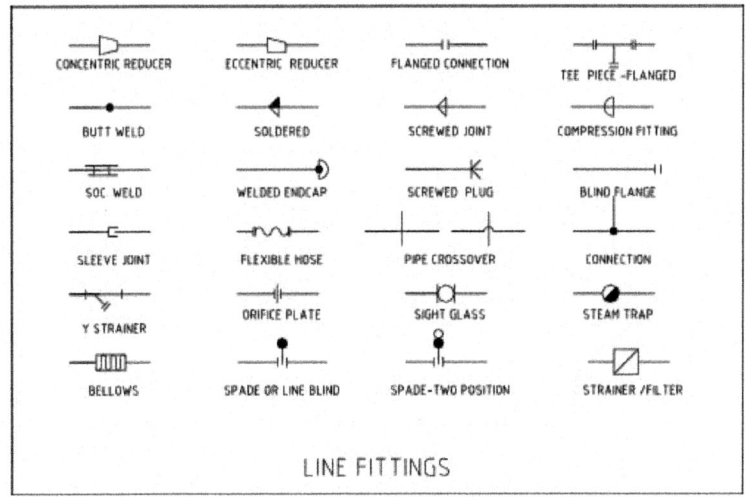

LINE FITTINGS

Table: Pipe Supports.

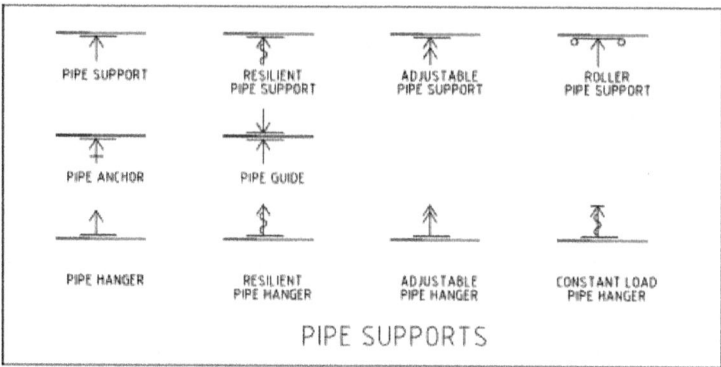

PIPE SUPPORTS

Crafting a P&ID

In order to greatly simplify P&ID diagrams for the purposes of this class, a standard convention must be employed. This convention simplifies the many control devices that need to be used. For the sake of brevity, sensors, transmitters, indicators, and controllers will all be labeled on a P&ID as a controller. The type of controller specified (*i.e.* temperature or level) will depend on the variable one wished to control and not on the action needed to control it.

For instance, consider if one must control the temperature of fluid leaving a heat exchanger by changing the flow rate of cooling water. The actual variable to be controlled in this case is temperature, and the action taken to control this variable is changing a flow rate. In this case, a temperature controller will be represented schematically on the P&ID, not a flow controller. Adding this temperature controller to the P&ID also assumes that there is a temperature sensor, transmitter, and indicator also included in the process.

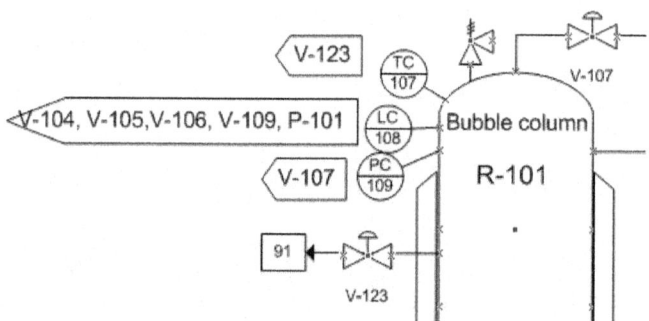

As you can see on the P&ID above, these controllers are represented as circles. Furthermore, each controller is defined by what it controls, which is listed within arrow boxes next to each controller. This simplifies the P&ID by allowing everyone the ability to interpret what each controller affects. Such P&IDs can be constructed in Microsoft Office Visio.

Sample Diagram

Below is a sample P&ID Diagram that is actually used in an industrial application. It is clearly more complicated than what has been detailed above, however, the symbols used throughout remain the same.

Table: Sample P&ID Diagram.

Figure B. Example System P&ID.

PID Standard Structure

A Piping & Instrumentation Diagram (P&ID) is a schematic layout of a plant that displays the units to be used, the pipes connecting these units, and the sensors and control valves. Standard structures located on a P&ID include storage tanks, surge tanks, pumps, heat exchangers, reactors, and distillation columns. The latter three comprise most of the chemical process industry single-unit control problems.

P&IDs have a number of important uses in the design and successful operation of chemical process plants. Once a process flow diagram is made, P&IDs help engineers develop control strategies that ensure production targets are met while meeting all safety and environmental standards. These diagrams are also very important for locating valves and process components during maintenance and troubleshooting.

Each page of a P&ID should be easy to read and correspond to a specific action of the plant. Also, symbols used in the P&ID are uniform throughout.

General Strategies for Implementation of Good Control Systems

There are several guidelines to follow when designing P&IDs for a plant. The first consideration is stable operation of the plant so that all safety and environmental standards are met. Maintaining product quality should also be a primary design objective. Additionally, systems should be designed to respond quickly to rapid changes in rate and product quality. Usually control systems should be run in an automatic mode so the system will correct itself, as opposed to a manual mode, which requires operator supervision.

When setting up a control system, it is useful to first focus on mass balance control, which can be monitored by level and pressure control loops that use gas or liquid flow rates. Additionally, a product control structure should be set up to ensure efficient process operation. To prevent controls from conflicting with one another, a final examination of the controlled streams should be performed. Further instrumentation can then be installed to prevent conflicts between controllers. Finally, appropriate tolerances should be established for controllers that directly affect the action of other controllers. Determination of the optimal control placement is essential for successful plant operation.

Standard Structures and Location of Control Features

As mentioned above, control of heat exchangers, reactors, and distillation columns represent the majority of single-unit control problems. The table below summarizes the control schemes derived in the following examples. The most common control schemes are listed in the table below.

Table: Summary of common P&ID structures.

Structure	Control Variable	Manipulated Variable	Location of Control Valve	Comments	Typical Disturbances
Steam-heated Heat Exchanger	Temperature of outgoing stream	Steam pressure	Steam line or condensate stream	Placement of a control valve on the steam line gives tighter control, but this type of valve is bigger and more expensive than one placed on a condensate line	Fluctuation of the inlet flow rate
Liquid/Liquid Heat exchanger	Temperature of outgoing stream	Bypass flow rate	Bypass stream	This allows high coolant flow rates and decreases fouling. Changes in bypass stream's flow-rate result in temperature control loop disturbances that are easily regulated.	Fluctuations in inlet flow rates
Jacketed CSTR w/ Endothermic RXN	Product stream temperature	Steam flow rate or steam pressure in the jacket	Steam line	Changing steam flow rates allows direct measure of the amount of heat entering the system and therefore a measure of conversion. It also makes a reactor sensitive to heat load changes and enthalpy changes. Manipulating jacket pressure is better at absorbing reactor heat duty and steam enthalpy upsets but it does not directly measure heat load on the reactor.	Fluctuations in feed flow rate, temperature, and enthalpy of heating medium
Jacketed CSTR w/ Exothermic RXN	Inlet temperature of coolant	Coolant flow rate	Coolant feed stream	There is a slow response to fouling on heat transfer surfaces, but there is a fast response to changes in inlet coolant temperature and coolant supply pressure.	Changes in feed flow rate, temperature; Fluctuation of coolant enthalpy

For most equipment, a degrees of freedom analysis is first performed and then control scheme is designed based on the degrees of freedom.

Heat Exchanger

To monitor the performance of a heat exchanger, the product stream is important. Usually, the product stream must be within some temperature range before it continues to downstream process units. The outlet temperature of this stream can be used to calculate the heat transfer. The steam is controlled in order to obtain the desired product stream temperature. One way to influence the product temperature is by controlling the flow of the heated steam. This flow-based control may take some time to implement and therefore cause fluctuations in the process. Depending on the process, these fluctuations may or may not be acceptable.

Instead of monitoring the flow-rate, the steam pressure may be monitored, achieving tighter control of temperature. A change in pressure is much easier to monitor and correlates directly with a change in temperature of the steam. This offers an effective way to control process temperatures. Pressure control also enables the physical condition of the piping to be monitored, since pressure changes occur as fouling progresses.

The pressure, combined with the flow-rate and temperature of the heated steam, can be used to calculate this fouling occurring inside a piece of equipment. It is important to remember that the steam should always be controlled on the inlet side of the heat exchanger for better pressure control and safety reasons, as shown in figure below.

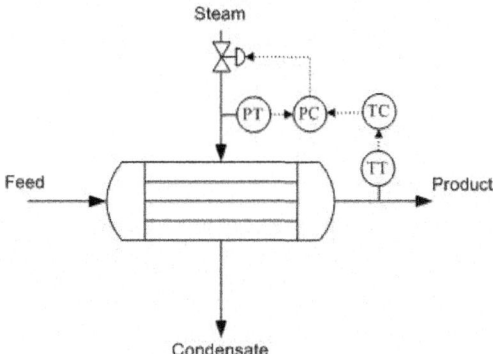

Fig. : Heat Exchanger with pressure control on steam inlet and temperature control on the product stream.

Distillation Column

Because the economic viability of an overall process is based significantly on product purity, it is important that distillation columns maintain stable operation. Changes in composition and flow-rate of the feed stream are common disturbances in distillation column operation. Improper functioning of controllers can undermine the effectiveness of the product composition. A degrees of freedom analysis can help place sensors and actuators in appropriate places, while not including too many sensors and actuators, in order to obtain an efficient control system.

Degrees of Freedom Analysis

A simple degrees of freedom analysis can make design of a control scheme easier as well as improve the control scheme overall. A process control approach to degrees of freedom, adapted from ECOSSE Module 3.1, requires tabulation of streams and extra phases.

$$DOF = Streams - ExtraPhases + 1$$

A typical distillation column contains the following streams; feed (1), bottoms (2), distillate (3), reflux (4), product (5), vapor-liquid mix (6), cooling water (7), and steam (8). There are also three locations where there are two phases, vapor and liquid, present in equilibrium. These are denoted as one "extra phase", since temperature and pressure are not independent in a two-phase system.

$$DOF = 8 - 3 + 1 = 6$$

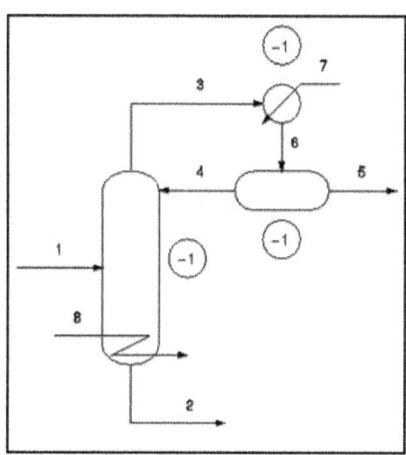

Fig. : Degrees of freedom analysis on a typical distillation column (adapted from ECOSSE).

In typical practice, the condenser and reboiler pressure are specified as atmospheric pressure. Once the pressure is specified in the condenser and reboiler(both two-phase systems), the temperature is specified and therefore the degrees of freedom are reduced by two. The total degrees of freedom, in practice, is six.

$$DOF = 8 - 5 + 1 = 4$$

A simple degrees of freedom analysis, in any system, can help define the variables and reveal where the critical sensors for process control should be located. This type of analysis also minimizes the design of too many sensors, actuators, and valves, which may in theory control and regulate the same variable.

Sensor and Actuator Options

Product streams, reboiler steam, and the reflux stream should have flow sensor/transmitters so that each flow can be adjusted to meet the column's control objectives. A flow sensor should also be placed on the feed stream because disturbance to the feed flow rate are common. Each flow sensor should be connected to a flow valve to control the corresponding stream.

Differential pressure level sensor/transmitters should also be used for the accumulator (the vessel that collects condensed distillate) and bottom of the column because maintaining these levels is essential for reliable operation of the column. If flooding is an issue, a pressure differential across the column should indicate the onset of flooding. A pressure indicator should be installed at the top of the accumulator to monitor column pressure. Temperature in a distillation column is typically controlled by manipulating steam flow to the reboiler.

Control Schemes

There are a number of common control schemes for distillation columns. Optimally, a distillation column should be run with *dual composition* control because it saves energy. In dual compostion control, the temperature of both chemicals in a binary distillation is controlled. The system is more complex to setup and measurements required for control may be difficult. As a result, many distillation columns use single composition control instead. Common control configurations for distillation columns include reflux-boilup and distillate-boilup.

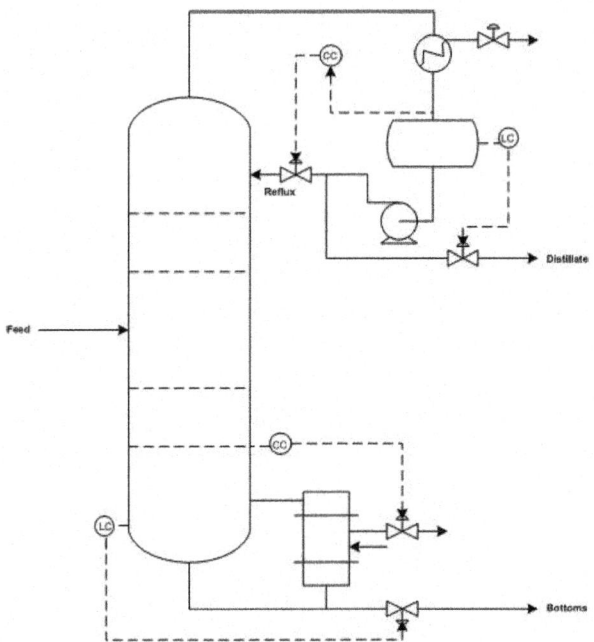

Fig. : Distillation column with reflux-boilup control scheme (adapted from Luyben).

In the reflux-boilup configuration, the distillate composition and bottoms composition are the control variables. The reflux flow and the heat input control (vapor boilup) are the manipulated variables, which allow control of the liquid and vapor flow-rates in the column. With this control system, a quick response to changes in the feed composition is possible. In the distillate-boilup control configuration, the distillate flow and the vapor boilup are used to control composition. This configuration is a better choice for columns where the reflux ratio is high.

Reactor (Exothermic Reaction in CSTR)

In implementing controls in reactive systems, temperature is a good indicator for unit performance. Temperature is often related to reaction rate and varies with time in most reactors. However, a reactor exit stream with a constant temperature is often desired downstream in the process.

In an exothermic reaction in a CSTR, lower initial temperatures result in lower reaction rates and low heat generation. As the reaction progresses, heat generation increases rapidly due to higher reaction rates and high concentrations of reactants. As reactant concentration decreases, the heat generation once again becomes low. Exothermic reactor temperatures must be controlled to assure stable reactor operation.

To remove heat from an exothermic reaction, basic heat transfer principles are employed. A coolant is pumped through a shell outside of the reactor. Since the heat removal is linear, the temperature of the coolant should be controlled. In doing this, it is possible to increase the driving force for heat transfer to slow a reaction, or conversely, to allow the reaction to further progress by decreasing the driving force for heat transfer.

Degrees of Freedom Analysis

In order to determine where to place controls, sensors, and valves in a exothermic reaction operation, a degrees of freedom analysis, similar to that carried out in the distillation section 3.2, can be helpful.

A typical CSTR contains the following streams; reactant A (1), reactant B (2), product (3), and coolant to the jacket (4). There are no locations where there are two phases, vapor and liquid, present in equilibrium - assuming the reaction is liquid phase with no simultaneous evaporation or sublimation.

DOF = Streams − ExtraPhases + 1

DOF = 4 − 0 + 1 = 5

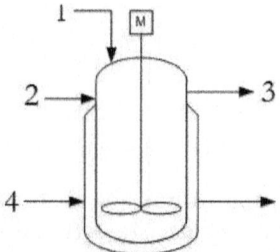

Fig. : Degrees of freedom analysis on a typical exothermic reaction in a CSTR.

Since in most practical applications the reactants (1 and 2) as well as the product (3) flow rates are defined by demand for the product, there are only 2 degrees of freedom which allow for placement of a 2 valves or controllers on the coolant stream (4) as well as one of the reactant streams (1 or 2). Only one of the reactant streams need to be specified as the other can be determined by a ratio controller using stoichiometric coefficients.

Control Schemes

As discussed previously, a valve can be placed on the coolant inlet stream to ensure proper temperature control of the reactor. By controlling the coolant stream based on inlet conditions of the reactant streams, the control can respond quickly. But, if the coolant is controlled based on outlet conditions of the product stream, there is a lag in response, but it is easy to monitor heat transfer performance.

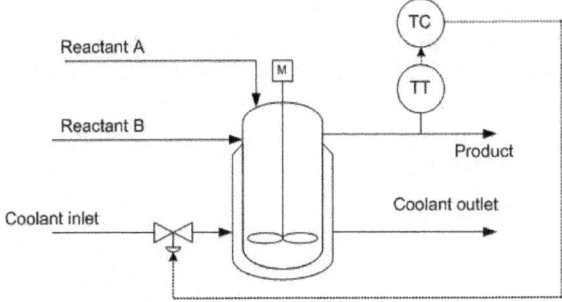

Fig. : Jacketed CSTR with coolant control based on the outlet temperature conditions.

To translate this example to an endothermic reaction, the coolant flow would simply be translated to steam flow in order to provide heat to the reaction through the CSTR jacket.

Other Common Process Equipment

Beyond the heat exchangers, reactors, and distillation columns, many other pieces of process equipment, including furnaces, compressors, decanters, refrigerators, liquid-liquid extractors (LLEs), and evaporators, are subject to disturbances and require careful control.

Furnaces

For example, furnaces may be subject to frequent load changes as a process or customer requires more energy. To cope with these demands, the temperature of the outlet stream must be monitored and manipulated. Information from a temperature controller at the outlet stream can be used to effect changes in valves that control the flow-rate of air and fuel to the furnace.

At this point, the best setup of the control system must be considered in light of factors such as safety, lag time for changes to occur, and equipment wear. For a furnace, the controls should generally be set up so that an excess of air is more likely than an excess of fuel. This reduces the risk of explosion due to the furnace filling with uncombusted fuel.

Liquid-Liquid Extractors

In liquid-liquid extractors, the interface level and pressure are the controlled variables. Disturbances in flow rate of the entering stream can affect interface level and prevent complete separation of the heavy and light components. From this, it is obvious that there should be valve controls on both exit streams. The best control scheme depends on the operation of the process. When the heavy phase is continuous (light phase flows upward through heavy phase), changes in interface level should be controlled by adjusting the flow-rate of the light product, while the pressure is controlled by adjusting the flow-rate of the heavy product out of the column.

Figure is a representation of what occurs in a single stage extractor. Generally, single stage extractors are using in chemical labs, where as multistage extractors are used in industry. A multistage extractor uses the immiscible liquid stream from the previous stage as the feed in the following stage. Figure depicts the control scheme previously described.

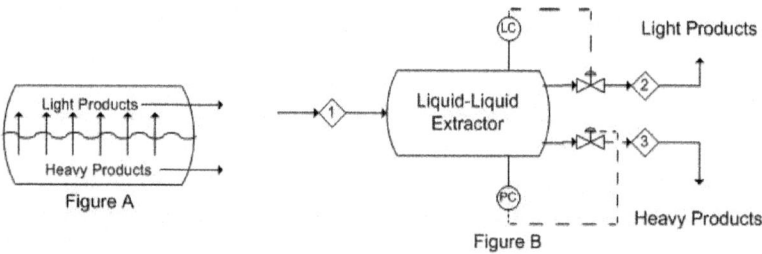

Fig. : Liquid-liquid extractor control scheme with interface level and pressure control.

When the light phase is continuous, the control system must be set up in the opposite manner. Figure is again a depiction of what occurs in a single stage extractor. Figure is a representation of a control scheme that could be implemented. This is the reverse of the control scheme in Figure.

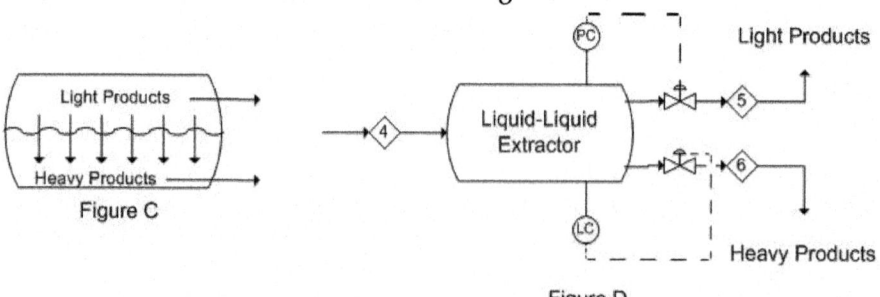

Fig. : Liquid-liquid extractor control scheme with light-phase control.

These representations are only two possibilities for basic control schemes that can be implemented in a process as there are several controllers and aspects of the specific processes that may need to be controlled or monitored.

Compressors

Compressors are another valuable component in process design. They allow for the reduction of the volume of an incoming stream by increasing the pressure at which the stream is maintained. They can also be used to pump liquids through the process, as liquids are highly incompressible compressors cannot be used for volume reduction.

For this, there must be a specific control system as to prevent adverse effects due to extremely high or low pressures. There are several types of compressors, including: dynamic, axial, and rotary [1] to name a few. Because the increase in pressure is governed by the ideal gas law, there is most often and increase in temperature as well. This can be left as is, or sent to a heat exchanger for temperature reduction. Heat exchangers were discussed above.

One such example is the use of a centrifugal compressor to reduce the volume of a fuel stream for storage. Using a compressor will allow for volume reduction as gasses are easily compressed, this is also economically friendly as it reduces the size of tank necessary to store the fuel stream. The tank should also be equipped with a pressure reducing valve, to bring the stream back to a desired pressure, depending on the process. A diagram of this scheme is as follows:

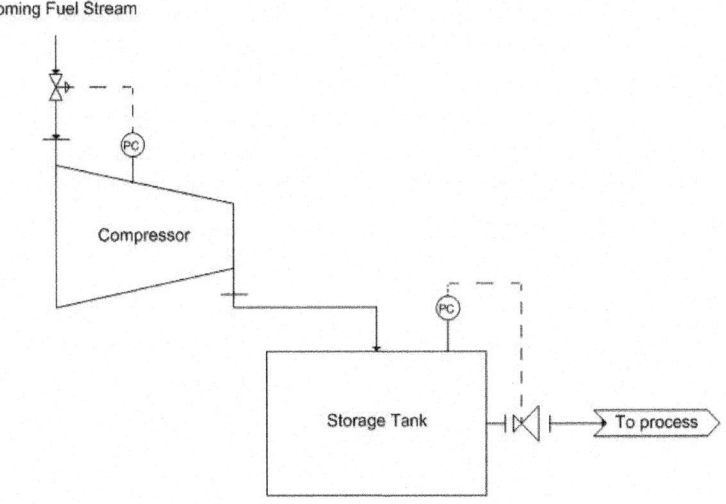

Fig. : Compressor control scheme with pressure control.

The pressure controller on the compressor controls the valve on the incoming fuel stream. This ensures that if there is a build up in pressure, the flow into the system will be stopped in time. Also, a pressure controller should be placed on the storage tank. This is controlled by the pressure reducing valve mentioned earlier.

Decanters

Decanters, much like Liquid-liquid extractors, use solubility as their principle of separation. Unlike Liquid-liquid extraction, these require some time for the separation to occur. Generally, the separation is a liquid-liquid or liquid-solid separation. Decanters are widely used in the wine industry to separate out sediment from the wine. Utilizing the wine separation example, a possible control scheme is as follows.

Here, there is only a level sensor on the decanter as it is a liquid-solid separation. Note that there is a release stream, used to remove collected sediment. An analytical or pH sensor could also be used here to maintain the correct chemistry or pH, as wine is the final product. It is also important to note that the exact placement of the level sensor would vary depending on the desired level in the tank. Here, the level sensor is not shown in its exact placement.

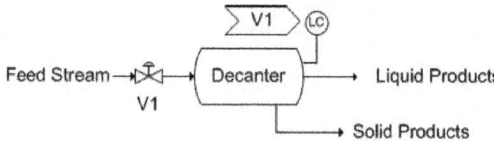

Fig. : Decanter control scheme with level control and pH control.

Another vessel that is very similar to decanters are knockout drums. These vessels are generally located after heat exchangers or other pieces of equipment that result in a multiphase system. These vessels are used to separate the two phases, generally gas-and liquid separation. A possible control scheme is depicted below. The incoming stream is a liquid-gas mixture coming from a heat exchanger.

Thus, there is a pressure sensor on the knockout drum. A level controller could be used, but this is effectively measuring the same thing, so it has been omitted. Also, because it is coming from a heat exchanger, a temperature controller has been included to control the amount of cooling taking place in the heat exchanger.

The pressure controller (PC) controls V2, the stream into the tank. The temperature controller (TC) controls V1, the valve on the coolant stream to the heat exchanger.

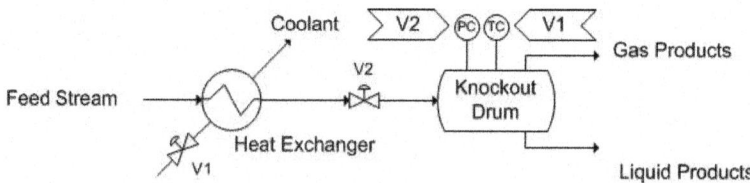

Fig. : Knock-out drum control scheme with pressure and temperature control on different streams.

These examples illustrate the typical method of locating control systems for process equipment. After choosing the location for valves based on process constraints, there still remain a number of possibilities for the actual manner in which to control the valves. The lag-time for changes may be longer for certain configurations, while the danger of damaging equipment is smaller. The controls configuration will depend strongly on the safety concerns associated with a specific process.

Selecting Controls and their Locations for a Multi-Unit Process

The following steps should be followed when setting up controls for multi-unit processes:

1. **Determine process objectives**, taking into consideration product specifications, economic constraints, environmental and safety regulations, *etc.*

2. **Identify boundaries for normal operation**. These can be based on equipment limitations, safety concerns, environmental regulations, and the economic objectives of the processes.

3. **Identify units and streams in the process that are susceptible to significant disturbances**. These disturbances commonly occur in feed streams, product streams, and reactor vessels, but can be present anywhere that temperature or pressure or other variables are changing.

4. **Select the types and locations of sensors** in order to properly measure and monitor critical process variables.

5. **Determine the appropriate types and locations for control valves** in order to appropriately adjust process variables so that they remain within the normal operating boundaries. Controls should be set up to minimize response time between sensing a change and taking corrective actions. The ideal location for any given control depends on the process unit or units that it affects.

6. **Perform a degree of freedom analysis.**

7. **Energy Considerations**. An energy balance should be performed for the process. This step involves transporting energy to and from process units. This may include removing heat generated by a reactor and using it elsewhere in the process. Control valves will help regulate the flow of such streams.

8. **Control Process Production Rate and Other Operating Parameters**. Adjusting process inputs, such as reactant feed rates, can alter other variables in the process. Process controls must be able to respond to these adjustments to keep the system within operating boundaries.

9. **Set up control system to handle disturbances and minimize their effects.**

10. **Monitor Component Balances**. Accumulation of materials within a system is not desirable and can lead to inefficiency in the process or catastrophic failure.

11. **Control individual unit operations**. Each unit of a multi-unit process needs to be individually controlled in order for control of the entire system to be possible.

12. **Optimize the process**. If the system has degrees of freedom, process variables can be manipulated in order to more efficiently or economically create product.

PID Standard Pitfalls

Piping and Instrumentation Diagrams (P&ID) are standardized in many ways, and there are some fundamental safety features that are absolute requirements for all P&IDs. Unfortunately, many people forget these features in their designs unintentionally. Lacking these safety features could lead to serious engineering problems. It is important to eliminate these pitfalls when designing a P&ID.

In the following sections, different pitfalls of P&IDs will be discussed. The equipment design section will concentrate on how equipment might be accidentally neglected and misplaced while designing a safe and functional process. The design process section will describe how a lack of consideration of process conditions would lead to serious design pitfalls. Overspecification and underspecification of equipment and design will be discussed. There are also some miscellanous pitfalls associated with interpretation and inherent problems of the P&ID itself. Examples are introduced at the end to illustrate common errors and pitfalls of P&IDs.

P&ID PRODUCTION PITFALLS

Most of the common pitfalls of P&ID production result from the engineer forgetting that the design on paper represents a real physical process and that there are practical and physical limitations that need to be considered in the process design. The following are some errors to be particularly careful of when drawing up a P&ID:

Equipment Design

Safety Valves

Safety valves are part of the essential valves system for P&IDs. Together with isolation valves, they are an absolute requirement for instrument design. Safety valves are required to install for all gas, steam, air and liquid tanks regardless of the tank's function for pressure relief purposes. Engineers should be aware of their system's set pressure, relief pressure, percent overpressure, maximum allowable pressures *etc.* when selecting a safety valve. The US law requires all tanks of pressure greater than 3 psig to have safety valves installed.

Different pressure tanks require different safety valves to best fit their safety design. Therefore, engineers must be very careful in selecting the right safety valves for their systems. However, when constructing a P&ID, engineers sometimes forget adding safety valves to their design, and this could cause serious problems. For example, if one forgets to add a pressure relief valve or safety valve on a reaction tank of gas and liquid, the extra pressure accumulating would exceed the preset pressure limits for safety design. This could lead to a serious explosion!

A pressure relief valve is symbolized by two triangles orientated at 90 degrees to each other, as shown below.

Pressure Relief Valve

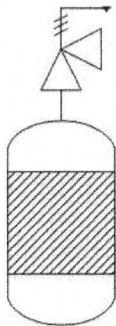

This eliminated excess pressure that might build up inside a reactor.

Isolation Valves

The isolation valve is used to isolate a portion from the system when inspection, repair or maintenance is required. Isolation valves are placed around the junctions in the distribution system. They are also part of the absolute requirement for P&ID construction. Engineers should be aware of their system's pressure, voltage, process medium, pipe sizing and flow rates when selecting the right isolation valves that work under the conditions for their system.

If engineers forget to add isolation valves on their P&ID design, serious problems would occur when the system needs to be partially shut down for maintenance or other reasons. There would be no way to control the unit's operation other than existing flow valves. Some upstream production problems could affect downstream production since appropriate isolation valves are not installed for safety and production purposes.

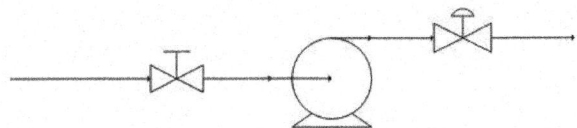

Isolation Valve Placement

Notice that the automatic valve after the pump is a flow valve, while the manual valve before the pump is an isolation valve. One example of this use is for cleaning. If some parts of the pump cannot be cleaned in place, the pump will need to be taken apart. The isolation valve can cut off the flow to the pump, allowing it to be safely taken apart.

Valves and Pumps

Some of the most commonly used pieces of process equipment that show up on P&IDs are valves. Valves control the flow of fluid through pipes by opening to allow flow or closing to stop flow. One of the problems associated with valves on P&IDs is the sheer number of them and deciding where to place them in the process.

A common mistake with valve placement has to do with the interaction between valves and pumps. It is important that valves be placed **after** pumps in a given pipe. When a valve is placed before a pump, should the valve close and the pump has not been shut off, there will not be a constant supply of fluid to pump. This is known as *starving* the pump. Starving the pump can create a disturbance known as cavitation, and it is very hard on the equipment and can cause the pump to break. Placing valves after the pump ensure that even if the pump does not shut off, it is still filled with the proper fluid.

Improper Valve Placement   Proper Valve Placement

These same principles apply to valve placement with respect to compressors. Placing a control valve upstream of a compressor eliminates your ability to control pressure in the pipeline downstream of the compressor and introduces a risk of starving the compressor.

Agitators

A point that is very easy to miss, and very expensive if it is missed, is that if a vessel is equipped with an agitator, the vessel **must** be filled enough to submerge the agitator before the motor is turned on. Agitators are designed to operate against the resistance of fluid. Without this resistance, the agitator will rotate much faster than the equipment is designed for.

This rotation may result in hitting the harmonic frequency of the structure, compromising its integrity and causing the agitator to rip from its foundation. This creates not only a fiscal predicament (not only ruining the agitator but also the vessel), but a safety nightmare. When designing a process, one must make sure he or she knows and accounts for how much fluid must be in an agitated vessel to avoid this situation. This can easily be solved by adding a level sensor to the tank that the agitator is being used in.

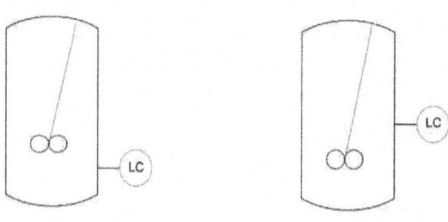

Improper Placement   Proper Placement

When placing the level sensor on the tank, make sure to place the sensor above the level of the agitator. This will ensure that the agitator is submerged in the fluid. It would be incorrect to place the level sensor below the agitator.

Instrument Selection and Placement

Instruments are designed to operate properly under specific conditions. Every instrument has a range over which it functions properly, and instruments must be selected that are appropriate for their applications. For example, a pressure gauge might have a working range of 5 psig - 50 psig. You would not want to use this gauge for sensitive measurements in the range 3 - 6 psig. Instrument material must also be considered during the selection process. If the substance being monitored is corrosive, for example, the instrument must be made of a corrosion-resistant material. Once an appropriate instrument has been selected, it must be appropriately placed. For example, a level control is not useful in a pipe because there is no need to measure any water level inside of a pipe, much like a flow controller is not useful in a storage tank because there is no flow.

Similarly, a flow controller should not be placed on a valve, but instead downsteam from the valve. However, level controls **are** useful in storage tanks and reactors, while flow controllers are useful in pipelines. Instruments must be selected and placed to reliably provide useful information and to accurately control the process.

Equipment Selection

When creating a P&ID, the equipment that is selected to be used is very important, not only to maintain a smooth process but also for safety purposes. Each and every piece of equipment from 100,000 liter storage tanks to temperature sensors has *Operational Limitations*. These are the conditions under which a given piece of equipment operates as expected, with safe, consistent, and reproducible results.

For example, when storing a highly pressurized gas at 2,000 psig, one wouldn't want to use a storage tank that has been pressure tested up to 3 psig. The process conditions are way outside the operational limitations and would pose a serious safety hazard. The same goes for sensors and gauges of all types. On a pressurized vessel at 2,000 psig, it would be no good to use a pressure control system that has a sensor that is meant to measure up to 100 psig, the results would not be accurate.

Operational limitations can usually be found in the equipment manual sent by the manufacturer or possibly on the manufacturer's website. If it is not found in either place, an engineer is obligated to contact the manufacturer and find the operational limitations before using a piece of equipment in a process.

Process Design

Unit Operation Input/Output

Providing clearly specified inputs and outputs to and from process units is vital. For the safety of the system, proper control and the prevention of disaster, it is important to show where each substance came from and where it is going. The P&ID must show all material streams to and from separation units, heat exchangers, and reactors.

For example, if the P&ID is not clearly denoted in a reaction to create construction explosive materials, then a large and perhaps fatal calamity can occur at the chemical plant. The exiting streams of a plant may pour into a river which may violate environmental regulations. Therefore, specifying inputs and outputs is imperative in a P&ID.

Pressure and Flow

The movement of fluid is essential in many production lines. Transporting material from a tank to a reactor, a reactor to a distillation column or from a column into a tanker truck all involve the movement of fluids. When designing P&IDs, a process engineer must decide how they are going to attack this problem, namely whether or not a pump is needed to move the fluid.

The main issue here is pressure. Fluid, both gaseous and liquid, moves down the pressure gradient from high pressure to low pressure. The rule of thumb is that if the source of the fluid has a much higher pressure than the destination of the fluid, a pump is not needed. Even if the source has a pressure only a fraction higher than that of the destination, a pump may yet be needed for the fluid to flow through the pipes, from the source to the destination.

However, the flow and pressure from the source must also be compared to the needs of the process. For example, if a inlet flow has highly fluctuating pressure, a pump or valve should be used to regulate the pressure for the process. Also, if the diameter of the pipe does not remain constant and is fluctuating as well, perhaps due to outside pressure or force, a pump should be used to control the pressure inside the system.

A common mistake young process engineers make is when charging a pressurized vessel, they do not use a pump powerful enough to overcome that pressure. This causes *backflow* and can ruin process equipment. This problem is especially prevalent in recycle streams. In some situations it may be appropriate to use a check valve (a valve allowing fluid to flow through into it from only one direction), to add an additional barrier against backflow. An example of such a situation would be when a fuel is to be mixed with air for combustion: if any air were to backflow to the fuel source, a dangerous, explosive situation would arise. In this case the use of a check valve would be appropriate.

An example of when a pump is never needed is when liquid from one vessel at ambient pressure is being transported to another vessel at ambient pressure that is at a lower elevation. This is known as a *gravity feed* and utilizing gravity feeds where possible can significantly decrease the cost of a process. As long as the pressure at the exit of the pipe is lower than the pressure created by gravity, a gravity feed can be used. An example of such a gravity feed is the flushing of a toilet, by which the water from the cistern at a higher elevation falls to the water closet at the lower elevation.

Under Specification

For safety and control purposes, **redundancy** is desirable in control systems. A process lacking in suitable redundancy would be **underspecified**. For example, a reactor containing an exothermic reaction would be underspecified if it only contained a temperature control. The potential danger posed by an exothermic reaction warrents a high degree of safety: if one controller were to malfunction, a backup must be in place. A pressure control should also be included, and perhaps a level control as well depending on the specific process.

Over Specification

On the flipside of under specification is **over specification**. Adding too many controllers and valves on vessels and lines is unnecessary and costly. Since process control is highly dependent upon the situation, it is difficult to set specific limits on the necessary amount of specification. However, a controller should not be used unless there is a specific need for it.

Misused redundancy is a common example of overspecification. It is unlikely that a water storage tank needs level, temperature, and pressure controllers. A degree of freedom analysis, as shown in a distillation column example, can be helpful in determining the exact number of controllers, sensors, actuators, and valves that will be necessary to adequately control a process.

Problems When Using a P&ID

Other than equipment and production pitfalls, there are also some general P&ID interpration and inherent problems.

Interpretation of P&IDs

Although it is essential for P&IDs to represent the right instruments with the right references, many P&IDs do not support scaling or do not require scaling as part of the system. P&IDs are drawn in a way that equipment and piping are displayed for ease of interpretation. They do not show the physical placement and location of different systems or the actual sizes and length of equipment and pipelines.

Experience tells us that many engineers have overlooked certain pieces of equipment or over-estimated the size of equipment and piping, leading to calculation and construction errors. For example, an engineer could be designing new equipment on the existing P&ID and not realize that existing equipment is so densely populated that extra equipment will not fit. It is important to check the physical space of an area before adding equipment to a process.

Inherent Problems of P&IDs

There are a few problems with P&IDs that are only solved by being conscious of them. One problem is that P&IDs are constantly being updated, revised, changed, and added to. When reading a P&ID it is always important to check the date it was last revised and if there is a later revision available. Making adjustments to equipment that isn't there anymore not only causes confusion and frustration, but is a waste of everyone's time. Moreover, making sure you are using the most recent edition to the P&ID will ensure that you do not purchase equipment to install that has already been installed.

Usually, companies will have a computer database with the most current P&IDs. Before modifying or working with an old paper copy of a P&ID, check to make sure it is up to date with the most current revisions.

Another problem with P&IDs is that even if they are well made and technically perfect, complex processes often appear cluttered. There is an inordinate amount of information contained in a P&ID, and as such P&IDs are next to impossible to take in at a glance. It is important to carefully study each document to fully understand the process. The mistake of assuming you gained all the information off of a P&ID from simply scanning it can leave you with mental holes that you don't even realize you have. This can be avoided by reading the P&ID carefully on the first pass and constantly referring back to it when questions arise.

Chapter 7

Process and Instrumentation Diagrams Safety Features

Safety has become integral to the manufacturing world. The implementation of proper safety techniques and accident prevention can not only save time and money, but prevent personal injury as well. P&IDs, when properly utilized, are a powerful resource to identify safety hazards within the plant operations. The following chapter provide an overview for the safety hazards that exist within a process, and illustrate the importance of P&IDs in a chemical plant.

SAFETY IN DESIGN

During the early stages of plant design it is critical to determine important safety features that remove potential hazards from effecting the facility environment. Regulations require that plant designers play a major role in minimizing the risks associated with these hazards. However, in order to do so, designers need to be aware of the hazards that exist during plant activity. The facility design team must develop a detailed drawing (P&IDs) including specifications of the plant process and environment to ensure that every aspect with regards to safety is covered.

Hazard *vs* Risk

When discussing safety, the terms hazard and risk are often used interchangeably. However, the difference in definition between the two terms is critical in utilizing the information they provide in increasing the safety within a plant. Hazard is defined as a potential source of danger and risk is defined as the level of threat associated with the hazard. A risk of a hazard occurring can be represented mathematically by the following equation:

$$Risk = Frequency \times Consequences$$

Frequency represents the probability that a hazard will occur, and consequence represents the impact of that hazard. Values for each parameter of the equation above are assigned by using either experimental information or educated judgment based on engineering models and experience. Each plant design process will have specific safety hazards and risks associated with it. Therefore, there is no predetermined value that can be assigned to each variable.

For example, in one situation, a water tower may be well supported inside the plant facility, and in another situation a water tower of similar structure may be hoisted against a rusted frame work outside of the plant facility. Both of these situations have different levels of risk associated with them. Needless to say, the risks associated with a plant setup can be reduced substantially by minimizing the probability that a hazard will occur.

Hazard Locations and Risk Hotspots

Common hazard locations exist in any place containing large amounts of energy. The degree of danger is proportional to the amount of energy stored at that location. Risk can be directly linked to kinetic energy, potential energy, work, heat, enthalpy and internal energy sources. Kinetic energy, otherwise known as energy in motion, is present in any moving component. The component may be vibrating, rotating, or translating, and this motion causes the kinetic energy of that part to greatly increase.

Within industry, personal injuries and fatalities are common hazards associated with moving parts. Reactors and cooling towers placed in high locations contain potential energy. If a structural failure were to occur within a plant, these structural units and their contents could fall from a large elevation onto another processing system or a human being, releasing all the chemical contents stored inside.

Similarly, the stored work in springs and other devices can cause fatigue and wear on the mechanical system over time, and result in eventual machine failure. Heat released from a reaction within a chemical reactor can be rapid and fatal if not accounted for. The buildup of heat can cause serious consequences with runaway reactions and boiler explosions. The enthalpy and internal energy of a reaction typically are the cause of most runaway reactions and destructive fires in a plant.

Risk Hotspots

Uncontrolled chemical hotspots created within a chemical plant are a common source of hazard, besides the hazard locations pertaining to energy sources such as kinetic and potential energy. Risk hotspots mostly occurs in the piping system and associated valves of the system, joints, traps, and various other piping elements. Possible malfunction of the system due to structural corrosion can be triggered by the failure to maintain the piping systems efficiently and periodically.

Even if one of the valves in the system has corroded and is unable to function properly, the fluid flowing through the pipe might get trapped, and the resultant buildup of pressure in the pipe may cause major safety hazards including fatal ones such as an explosion. If the system is not shut off before the pressure gets out of control, the pressure buildup in the pipe will cause it to burst, releasing all of its internal contents to the surrounding environment.

Storage vessels are other pieces of equipment that must be cared for properly. Since the plant operators and engineers do not usually interact with the storage vessels, as compared to other parts of the plant process such as the piping system, they are considered to be of secondary importance and commonly overlooked. Storage vessels have much more content inside them than pipes, so a leak or a burst would take longer to get back under control than pipes, which can by plugged more easily. Problems which may arise from storage vessels are not associated with their design, but in fact from not thoroughly and periodically maintaining them. Possible complications that can arise from neglecting the storage tanks are over-pressurization, overfilling, heating element malfunction, or simply equipment malfunction.

Chemical reactors are another common location where risk hotspots occur. The nature and design of commercial chemical reactors is to handle a controlled explosion. However, if the control element is removed, disaster is bound to occur, just as in any other part of the chemical plant. The most common type of hazard is a runaway reaction inside a batch reactor. When the plant facility looses electricity or cooling water, a runaway reaction will strike inside the reactor.

Once a runaway reaction has spun up inside the reactor, many other hazards may follow, such as flow reversal in the pipes, incorrect reagent charging, heat exchanger failures, and external fires. Other hazards may perhaps be even more serious, such as engineering errors that could potentially cause a runaway reaction to occur, including inappropriate material selection, inadequate equipment inspection or failure to fully understand the chemistry or exothermic nature of a reaction. A runaway reaction originating inside the chemical reactor can easily cause a chain reaction across the rest of the equipment at the facility, and can result in the entire system malfunctioning.

Other process equipment that may be hazardous and where risk hotspots commonly arise, are vacuum operators, furnaces, pumps, gas movers, compressors, and heat exchangers. The location and type of specific piping and unit operations are available on the process P&ID. A responsible process engineer should use the P&ID to identify all risk hotspots, and act accordingly to monitor and maintain a safe working environment. In addition, a standardized plan should be constructed so that in the event of a malfunction, the correct steps can be taken to bring the faulty part back under control. Supplementary precautions should be taken to prevent a comparatively minor malfunction in the system from becoming a disaster which may violate environmental regulations and even endanger human lives.

Safe Design Principles

The ISD or Inherently Safer Design movement was a doctrine striving for safer chemical processing procedures. This movement was pioneered by Trevor Kletz in 1976, and promotes the design of processes so safe, that no catastrophic failure can occur within the plant. The following principles apply to initial process design:

1. Use the fewest number of hazardous substances in the smallest quantities and still maintaining plant productivity
2. When possible, substitute hazardous chemicals with chemicals that are less dangerous
3. Practice moderate operating conditions in the plant
4. Use the simplest plant design possible
5. Design equipment in the plant to minimize the effects of a hazardous incident

The specifications determined by the process designers are communicated through the P&ID.

Hazards in Construction

In order to eliminate hazards, a operations personnel must be able to identify that a hazard exists. Hazards that may be encountered on plant sites may be categorized into three main types:

A) Hazards harmful to health
B) Hazards likely to cause personal injury
C) Hazards likely to lead to catastrophic events

Hazards harmful to health:

When workers are exposed to or come in contact with asbestos, corrosives, irritants, toxins, or noxious gases try to avoid by specifying the processes, which lead to this exposure

Hazards likely to cause personal injury:

Hazard awareness is increased when people have to work in situations likely to expose them to the risk of personal injury, including moving plant machinery or working in areas where objects are likely to fall. Situations where there are live electrical circuits overhead, buried power lines, and confined working conditions are likely to cause personal injury.

Hazards likely to lead to a catastrophic event:

These hazards have consequences beyond the site boundary. They include fire outbreaks, explosions, flooding, or premature collapse of structures, cranes, tunnels and excavations.

Fail-Safe Design

The fail-safe design of a unit operation (such as a reaction vessel) requires a complete understanding of the operation at hand, and the knowledge of all the worst-case conditions. A fail-safe system is a unit operation such that, if any or all of the worst-case conditions were to occur, the operation would shut itself down automatically and in a safe fashion. In the case of a run-away reaction, if reagent feed limits, interlocking controls, and integrated heat balances are all properly maintained, the reaction cannot "run away." Other precautions such as purges, vents, dump tanks and quenches are available for reaction vessels, and should be visible on the P&ID.

Inherent Safer Predesign

The table below provides a guideline for identifying and minimizing hazards partly based upon Kletz's rules for ISD in the Safe Design Principles section above. The step/rule column describes the action taken. The tools column describes the mechanism by which the actions are taken. The experimental and analytical resources column describes the knowledge by which the mechanisms are created. The Literature References column describes where the knowledge can be found.

Step/Rule	Tools	Experimental and Analytical Resources
1. Identify	Fire triangle, Flammability-limit chart, Chemical reactivity chart, Safety compatibility chart, Safety stream chart, Toxicity ratings	Thermodynamic calculations, Reactor design equations, calorimetry, flammability charts
2. Eradicate (ISD Rule 2)	Inert-gas blanketing, Failsafe design	Thermodynamic calculations, Reactor design equations, calorimetry, flammability charts
3. Minimize, Simplify, Moderate, Attenuate (ISD Rules 1, 3, 5)	Moderate ignition sources, Keep reactive volumes small	Logical chemical engineering analytical and computational skills
4. Isolate (ISD Rule 4)	Separate hazardous operations - Surround in an impenetrable structure	Logical chemical engineering judgment

RELIEF SYSTEMS

Organizations such as the American Society of Mechanical Engineers, American Petroleum Institute and National Fire Protection Association layout recommendations and design standards so that most engineers with proper training can setup proper emergency relief systems for single phase flow. Unfortunately, this is often not good enough for reactive systems. These systems are much more complex and include multiple phase flow, runaway reaction potential and self reactant material.

When designing an emergency relief system (ERS) it is necessary to understand all aspects of the chemicals and processes that will be in play. This includes but is not limited to: kinetics of the possible reactions, contamination, interactions with air, rust, piping, or water, phase changes and runaway reactions. The following topics deal with the hazards that require designing relief systems and the prevention of runaway reactions.

Planning and Design

When designing an ERS, it is important to consider the worst-case scenario. This is based upon a thorough knowledge of the reactions, materials, and environment of the process. Some of the most critical scenarios are over-pressurization in a reactor.

All disaster scenarios can be analyzed using a hazard and operability (HAZOP) study. This HAZOP study will analyze a process based upon human, equipment, and environmental factors.

The HAZOP team must take the following steps to ensure that all potential scenarios are taken into account:

- Review the potential hazards of all chemicals. This includes non-operational conditions and interactions with contaminants
- Study the chemical process including all possible reactions, rearrangements, decompositions, *etc.*
- Review the P&ID's for the process
- Study the specific reactor and storage vessels for material composition compatibility, size, surface area, instrument ranges, and set points

With this information, the HAZOP team can determine most of the potential disaster scenarios. Most likely the worst case scenario will involve fire induced runaway reactions.

Design Strategies

There are various techniques often used to prevent failure from over-pressurization, fire, runaway reaction or other disaster scenarios. The following are some design suggestions that will decrease this risk:

- The use of insulation in case of fire. Unfortunately, insulation will also minimize heat loss from the reactor during a runaway reaction. It is important to consider the ability of a vessel to drain when using insulation. If the contents can be drained and cooled before the reaction starts to runaway, catastrophe can be prevented.
- Design to avoid fire damage on sensitive equipment. The possibility of fire damage on electrical equipment or sensors will cause control difficulties when trying to slow down a runaway reaction, or monitor another emergency situation.
- Consider the structural integrity of the entire system due to fire damage.
- Use multiple purge streams and valves to separate materials. Separating reaction components into a storage vessel with a large surface area will allow for faster cooling.
- Install two separate relief devices in case one has been compromised due to fouling or solid particle blockage.

Overall safety relief plans must be made for all possible scenarios.

Reaction Kinetics

Reaction kinetics usually determine the potential for safety disaster. For the average exothermic reaction, the reaction rate doubles with every 10°C increase. This will lead to an exponential increase in energy which will force both the temperature and pressure in a system to uncontrollable levels. Since reaction rates are sensitive not only to temperature, but also pressure, contaminants, concentrations and phases, all possibilities must be adequately analyzed for plant operation to be deemed safe.

Relief devices should be designed to handle chemicals and/or mixtures in any phase. For example, if a reactor temperature increases suddenly and a safety relief valve is activated the material must be able to flow through the valve quickly and safely. If the chemical has changed phases, or has a higher pressure that the valve cannot accommodate, the chemical will not be able to escape and build up pressure in a reactor. This would greatly increase the risk of a disaster. Because of the extent of possibilities when considering reaction rates and kinetics, it is usually necessary to consult some sort of computer simulation or dynamic simulation tool to plan for every possible scenario.

Runaway Reactions

Runaway reactions are caused when exothermic reactions are fed more energy due to malfunctioning cooling systems. This causes an exponential increase in temperature, which in turn causes an increase in pressure, and finally damage to the reactor and/or plant.

The possibility of malfunctioning cooling systems must always be considered for flammable materials. If materials are being stored at temperatures above their flash point, fire is always a possibility. Fire will cause a reaction to reach runaway conditions with very little reactant. Relief systems for fire induced runaway reactions must be larger than conventional runaway reactions.

While relief systems are often designed for over-pressurization, this might not be enough when considering flammable materials. Constant or prolonged exposure to flame will cause most normal reactant or storage vessels to fail causing chemical leaks or plant fires. Fire proof insulation must be used together with normal pressure relief systems to prevent system failure.

An important design variable when considering fire induced runaway reactions is fire flux. This variable considers heat impact on a reactor due to fire. Formulas for calculating fire flux can be obtained from the National Fire Protection Association. The elevation of flames must also be considered when designing reactor vessels and safety relief systems. Pool fires can produce flames that are hundreds of feet high; using the P&ID will be important in ensuring that process components sensitive to fire are sufficiently protected from something like pool fires.

Two Phase Flow

For systems in which multiple phase flow is possible, all phases must be planned for. For vapor hybrid systems, all-vapor flow should be considered. Likewise, when foam flow is possible, all-foam flow should be planned for. This is all part of the mindset of planning for the worst case scenario. The most conservative design should always be used when faced with uncertainties of phase flow.

Often times, a runaway reaction will cause high-viscosity two phase flow. Relief valves and safety features must be ready for viscous flow. Many resources suggest averaging the viscosities of the two phases to plan for two phase flow. It is important to keep in mind however, that two phase flow discharge will separate in the discharge line.

This will lead to higher pressure drop. Piping sizes are often underestimated due to this unplanned scenario. Undersized piping will lead to valve rupture and back pressure and could cause venting disturbances during the emergencies where venting is critical to the safe shutdown of a process.

Regulations

There are often conflicts and contradictions in federal regulations and recommended practices for safety design guidelines. When faced with such confusion the designer should at least design to the level of the Occupational Safety and Health Administration's (OSHA) requirements, as well as meeting regulations set by the Environmental Protection Agency (EPA), the Process Safety Management(PSM), and the Risk Management Program (RMP). While these guidelines and regulations produce a minimum standard to follow, when dealing with reactive systems, this does not always meet process needs.

Additional Safety Hazards in Chemical Plants

This section outlines potential safety hazards commonly found in places where chemicals are stored or chemical processes are taking place. While these hazards are rather easily prevented through attention to detail and general awareness, neglecting them can have catastrophic consequences.

Time Sensitive Chemicals

Some chemicals have a "shelf life," or an expiration date provided by the manufacturer. The chemical must be used by this date or properly discarded. These are typically reactive chemicals, which can become unstable after a certain period of time, possibly rupturing the vessel in which they are stored. An example of this is a monomer that begins to polymerize unless an inhibitor is present. This inhibitor is completely consumed after a certain period of time, allowing polymerization to occur, and therefore must be used or discarded by that time.

Another example deals with the formation of peroxide, which can be a severe fire and explosion hazard. It may also be a health concern, causing severe mucous membrane, respiratory tract, skin, and/or eye burns. Peroxide-forming materials should be stored carefully, labeled with the date received and the date first opened on the container. Chemicals should be disposed of or checked for peroxide formation after six months; do not open any container with solid formation around the lid. There are several ways to prevent this from happening.

Material Safety and Data Sheets (MSDS) should be available for any chemical the plant uses. These can inform you if a particular chemical becomes unstable after a certain period of time. It is also important to make sure there are procedures in place for handling time-sensitive materials. Lastly, investigating near-miss occurrences can help to ensure that future incidents do not occur.

Pressure Relief Systems

Any open pipe in a chemical plant is a potential discharge site. While operator convenience and maintenance remains a concern, safety takes precedence when dealing with relief systems. It is essential to operator safety that relief valve discharge sites are located in areas that pose a low risk of exposing personnel to chemical hazards and are directed away from all access platforms. It is common in industry to tie multiple pressure relief devices to an emergency vent header that releases on the roof of the plant. Long stretches of unsupported pipes also pose a potential threat. Force generated by material flow could bend or break the pipe impeding on plant operations, and more importantly, injuring personnel.

Dust Explosion Hazards

Most flammable solids can form an explosive dust cloud if the particles are small enough. Materials such as wood, grain, sugar, plastics, and many metals can all form these explosive dust clouds. Dust explosions occur when a combustible material accumulates in a confined area and is exposed to an ignition source. High risks areas are usually those that are neglected such as tops of vessels and tanks, on pipes, storage bins, bucket elevators, and dust collectors.

To prevent dust explosions, it is essential to implement good housekeeping practices. Care must be taken so that the cleaning process, such as sweeping, doesn't turn a dust layer into a more dangerous cloud. Operations and process engineers should be aware of all the mechanical and safety control equipment associated with preventing fires. Equipment like electric vacuums must be appropriate for use in an area where an explosive dust cloud could form. This entails no holes or cracks in the cord, sufficient grounding, and receiving site approval for use of the equipment.

Overfilling Tanks

The overfilling of vessels has long been a leading cause of serious incidents in chemical and petroleum industries. When a level sensor or high-level alarm fails, reactive material can spill over the tank and accumulate. If this material is exposed to an ignition source, there is the potential for an explosion, leading to serious property damage, environmental issues, and injury to operations personnel. When filling or draining a tank with material, operators should be aware of all relevant level, pressure and temperature controls in place, and watch for abnormal trends.

Also, all safety critical alarms surrounding a vessel should be tested regularly at frequencies recommended in plant process-safety-management procedures. Conducting regular process maintenance on safety critical alarms have numerous benefits, which include reducing the risk of operating a plant with faulty equipment, and increasing operator awareness with the location and function of critical safety devices.

Containing Storage Tank Spills

Engineering controls are implemented into plant designs to account for potential disasters, such as a spill, leak, or complete emptying of a storage tank. Industrial-sized chemical plants store large amounts of raw materials, products, and byproducts on-site. The amount of each varies, but is typically between three to thirty (or more) days of the required supply or amount produced. The stored chemicals can be hazardous, flammable, explosive, and/or reactive with each other.

In the event of a tank spilling, dikes are built around tanks to contain the spill and protect the surrounding community from the spill. The regulations of the dikes include the following: dike volume must be 1.5X the largest storage tank contained by the dike, reactive materials cannot be stored in the same dike, and scuba gear must be present on-site if any dike is deeper than four feet.

Since the dikes cannot contain any reactive materials, the implementation of dikes affects the overall plant layout. Typically, dikes are designed to be like speed bumps and have a height less than one foot so fork-lifts and tankers can easily maneuver through the plant. Dikes are an effective engineering control that greatly improve the safety of a chemical plant.

Temperature and Pressure Ratings

Before a vessel is put into plant operation, it is rated and stamped by the manufacturer with temperature and pressure limits. Problems arise when personnel overlook the inverse relationship between temperature and pressure for gases. Be aware that equipment rated for a specific temperature and pressure, cannot be operated at the same pressure if the temperature is increased.

While it may seem extremely intuitive (think ideal gas law), this relationship is too often overlooked, usually with serious consequences. When operating any process, pay attention to the temperature and pressure ratings. If they aren't readily known, review equipment files or contact the manufacturer before making any changes to the process.

Also, operation and maintenance should always be performed according to strict standards laid out by a plant standard operating procedure (SOP). Any changes must be reviewed and approved by a cross functional team.

The above hazards and preventive measures have been outlined to illustrate how important safety is in a chemical plant, and the importance of being aware of your surroundings and all possible safety hazards. Many of the aforementioned topics may seem like common sense, but it is very easy to overlook small details in the scope of a large-scale chemical process. Keeping safety in mind at all times as the paramount of any process can ensure that people leave their shift the same way they arrived.

Alarms in Processes

Alarm configuration and specification is an important part in the design and operation of any chemical process. Alarms are implemented in a process design to aid in the control of the process. Federal and industrial documents only specifically reference alarms in the context of processes exceeding regulatory compliance limits.

In this sense, alarms are used to control safety and environmental hazards. Other important uses of alarms are to control product yield, product quality, and operational limits of process equipment. This section will discuss the steps taken to implement alarms in processes, common different levels of alarms, and common instances in which alarms are useful to comply with regulations.

Alarm Lifecycle

When it has been determined that an alarm is needed to aid in the control of a process, the alarm must be specified. The following are basic steps in implementing an alarm.

1. First, the process designer needs to know what category the alarm fits into. This is important because responses from different categories of alarms are usually managed differently. They may be prioritized in case multiple alarms occur at once, so that the proper follow-up reports and procedures may be taken accordingly. Common categories include product quality parameters, safety, environmental considerations, and equipment protection.

2. After determining the specific use of the alarm, the limits must be set. If the process variable exceeds the set limits, the alarm will be triggered.

3. Next, the computer system for the alarm must be configured. The computer system may contain logic loops that automatically change control parameters to offset the problem and merely inform the operator that it has done so. Alternatively, an alarm may trigger horns, flashing lights, or send a page to an operator, alerting them that there is a problem that needs attention.

4. The correct user response and interaction must then be defined for the alarm. This includes providing proper training for the particular process, procedures and operator manuals describing how the event should be

investigated, guidelines on when action needs to be taken, and guidelines on when to escalate the situation to a more serious event.

Proper communication from the automated alarm system is critical. Care needs to be taken that alarms only signify abnormal conditions that require a response. For example, the successful completion of a batch operation is an important piece of information and should generate a computer message so that the operator knows. However, this does not represent an abnormal situation and should therefore not show up on the computer as an alarm.

While this seems quite obvious, most industrial plants struggle with maintaining alarm systems free of "nuisance alarms." It is also important that alarms have proper descriptions so that the reason an alarm appears is clear. For example, an alarm could appear in a large plant that says, "LI-501 exceeds limits" by default. Although this might be useful to the engineer designing the alarm system as an indication that tank 501 has a high level of material, an operator that sees the alarm or other engineers working on the system may not know what the alarm is communicating.

A more universally meaningful alarm indicator might say, "Material level in tank 501 is high." With this simple change in the computer system, the alarm would be more effective for personnel to locate the problem quickly and act accordingly.

Alarm Levels

Alarms are available in a wide variety of types, with multiple levels of alarm. In all processes, disturbances occur that can shift a plant's operation away from normal. When this happens, measures are usually taken by computers, such as with the use of P&ID control loops, to keep the process under control. With these control systems, processes are designed to fall within a range of acceptable normal operating limits. When a process deviates beyond these normal limits, an alarm should be triggered.

For most processes, the minimum for safe operation is two levels of alarms: warning and critical. The warning alarm tells plant operators that the process has deviated beyond the acceptable limits and provides them with the time and ability to take corrective action so that the product quality is not affected and environmental and safety regulations are not exceeded.

If the right actions are not taken or are not taken quickly enough to correct the problem, a critical alarm may then be triggered. The critical alarm tells the plant operators that conditions are dangerously close to breaching what is allowed. In many cases, the critical alarm will call for a systematic shut-down of the operation until the problems can be addressed.

The conditions at which warning and critical alarms are triggered are those conditions that exceed the limits determined for the process. Measurement uncertainty must always be considered because all devices in the control system will be subject to some possible error, even if it is small as +/- 1%. Setting an alarm at exactly the proven acceptable range for the process could allow a measured value to fall within this range, even though the actual value lies outside. This is called a "false acceptance." By performing error analysis and statistical distribution theory, the alarm limits can be adjusted as needed. This is a process called "guard banding," and it prevents real disturbances in the process from being ignored by the alarm system. Information about alarm limits should be well documented so if changes to the system are proposed, designers know how the limits were originally determined.

The figure below provides a visual representation of alarm ranges.

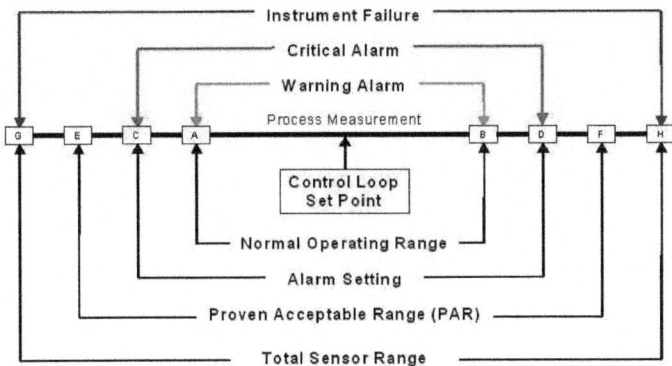

As seen in the center of the above figure, the control loop set point is the optimum point of control for the process (*e.g.* the optimum temperature and concentrations of reactants for a reaction). It is impossible to maintain the process at exactly this point, so there is a range of "normal" operation, inside which the process is still considered to be running in an acceptable way. The warning alarm would be triggered when the process goes outside of the limits of this range (lower than A or higher than B), allowing time for the process to be brought back under control.

The critical alarm would then go off if the process goes beyond the alarm setting (lower than C or higher than D). This setting is determined by guard banding the process acceptable range (PAR) for uncertainty, seen in the figure by the fact that the alarm setting lies well inside the PAR. The spaces between E and C and between D and F are determined by the uncertainty. Lastly, this PAR must be inside the total range of the sensor that determines the instrument failure.

Alarms must be analyzed based on their priority:

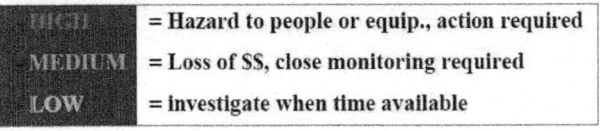

Safety Steps to High Reliability

A safe system involves many layers of reponses when an incident occurs.

EMERGENCY RESPONSE

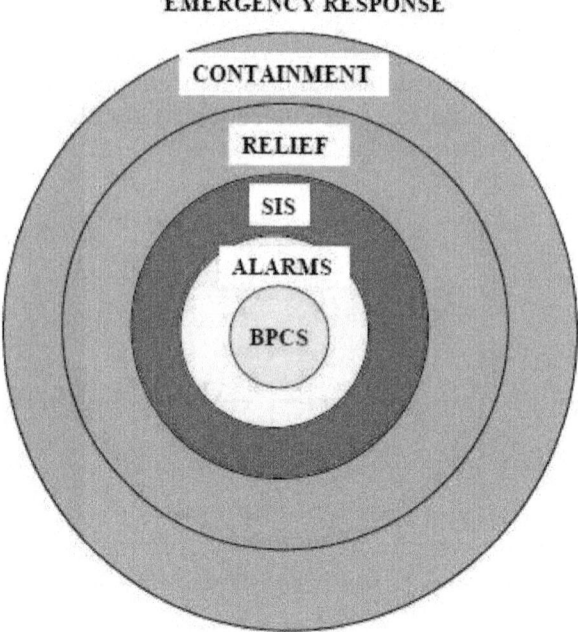

The center of the ring is the basic process control system.

The first layer of response is the alarm system which draws attention.

The second layer is the Safety Interlock System which can stop/start the equipment.

The third layer is the Relief system which leases pressure build-up in the sytem.

The fourth layer is containment which prevents material from reaching workers, community, or the environment.

The last layer to the ring is the emergency response system which involves evacuation, fire fighting, *etc.*

Worked out Example

A reagent recovery unit for a chemical process plant is being designed. The goal is to recover tin from a tin-plating waste stream through binary extraction with carbon tetrachloride. The equipment used in the process and P&ID are as follows:

- Carbon tetrachloride storage tank
- Mixing Vessel
- Pumps
- Heat Exchanger
- Associated piping

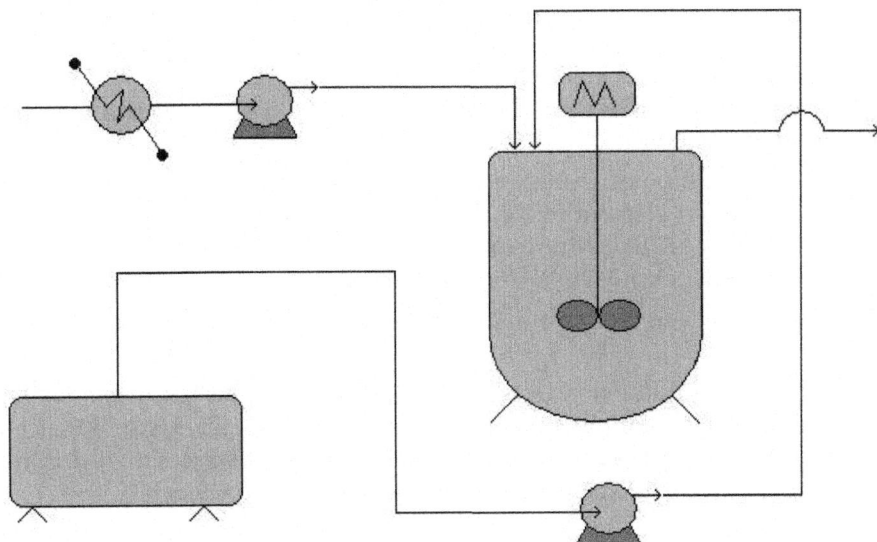

With knowledge of hazardous locations and risk hotspots, and safe design principles, analyze the flow diagram. Identify areas of risk and specific improvements that should be made to the process design before implementation begins. (note: no piping, valves, or controllers exist on the P&ID for the sake of clarity. Ignore these in your analysis).

Areas of Risk

Any piece of equipment could potentially become dangerous if the right situation were to arise. Each piece of equipment used in this process is no exception.

- Storage tank - These process units that are of secondary importance to the process goal, don't draw as much attention from plant operators and engineers, and tend to receive less maintenance.

- Mixing Vessel - The potential for heat buildup due to heat of mixing and the kinetic energy of the spinning motor makes this vessel a risk hotspot. Improper control of temperature within the vessel or motor speed (especially during periods when the tank may be empty) can lead to equipment malfunction or even explosions

- Pumps and Heat exchangers - The energy stored through pressurization by the pumps and potentially high temperature solutions handled by the heat exchanger present the possibility for danger. Wherever energy is stored, danger is associated with its potential release.

- Associated piping - Corrosion or failure to maintain pipes and associated elements are the main cause of the malfunction leading to danger. Leaks or total failures can release reactive materials on other equipment causing corrosion or malfunction

Specific Improvements

This process could be improved and made safer in the following ways.

*   Simplification of the piping used to connect the storage tank to the mixing vessel will reduce the risk of leaks and malfunctions due to piping complications.

*   The addition of a emergency relief valve and temperature controlling/insulating element to the mixing vessel will greatly reduce the risk of equipment failure due to overheating or overpressureization due to heats of mixing or outside heating influences.

*   Use a reactor with a rounded top, as this type can withstand much higher pressures.

*   The substitution of carbon tetrachloride with a less hazardous solvent, if the specific solubility required to extract the tin still exists, like cyclohexane, isopropyl alcohol, or 1,1,1-trichloroethane reduces the risk of health complications due to exposure, and possible explosions.

Example: Alarms in P&ID

A P&ID appears below for the production of a solution containing a pharmaceutically active compound. The reaction taking place in the CSTR is highly exothermic. After examining the P&ID for this part of the process, describe a possible alarm system.

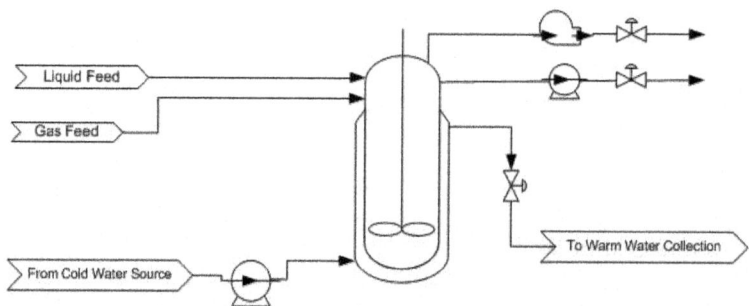

Answer: The CSTR for the exothermic reaction is jacketed with a cooling water stream. An alarm should be in place to monitor the reactor temperature. A warning alarm can notify the operator that the temperature is too high and corrective action needs to be taken. A critical alarm should be in place to warn that the reactor is nearing runaway conditions and an immediate response is needed. If the necessary action is not taken, systematic shutdown of the reactor could occur. This would involve closing the valves, flooding the jacket with cooling water, and having the impeller on. Another possibility for an alarm, although we do not know how the products are being used specifically, is in a composition measurement of the product containing the pharmaceutically active compound. Depending on where this stream is going and how it is being used, too high a concentration could be dangerous if no other concentration-altering steps occur before the finished product goes out to consumers.

REGULATORY AGENCIES AND COMPLIANCE

Regulatory agencies govern various sections of chemical plants from emissions and water treatment to personnel behaviors and company policy. Understanding of the types of organizations that demand high performance from chemical plants on a national and local level are extremely important for the safety of the employees, community and environment.

Regulatory Compliance

Compliance is an integral part of ensuring the safety of all that work in close contact with the chemical plant as well as minimizing fines and fees that come with violating regulations. Compliance can be organized into two main categories; plant safety and environmental safety. Plant safety ensures employees and the surrounding community are adequately protected during full-operation of the plant. Environmental safety comprises protecting and treating the environmental carefully when it comes to various contaminated streams in the plant.

Plant Safety for Employees and the Community

Safety is paramount in any chemical process, and alarms are essential for compliance with safety constraints. OSHA has established guidelines that must be followed in any plant when dealing with chemicals defined as highly hazardous. Limits for temperature, pressure, flow rates, and compositions need to be regulated. Alarms should be in place to warn operators when a limit is near so that steps can be taken to ensure the safety of people in the plant. An example of this is a safe temperature limit for a CSTR. The alarm signals that the temperature is too high and action is needed to prevent a runaway scenario. If the corrective action is not taken, or not taken quickly enough, a critical alarm can signal a computer program to automatically shut down the entire process or specific unit operation.

Some industries may expect product quality to be closely regulated by the FDA or other government agencies. Typically, this will be in a process where the final product is directly used by people and the margins for error are small. These processes include food processing and manufacturing consumer products, especially pharmaceuticals. These industries usually require systems in place that frequently validate alarms, as well as documentation for all critical alarm events. Measurements such as the weight percent of a pharmaceutically active compound in a solution must be carefully monitored, with recorded uncertainty analysis.

Environmental Safety

Emissions of solids, liquids, and gases in a plant are heavily regulated by government agencies. Regulations apply for processes that emit chemicals to the atmosphere (either directly or following a scrubber), processes that discharge material into a body of water, or processes that require containment control devices like check valves and rupture disks. Alarms are frequently used to comply with these regulations by measuring things such as pH and organic solvent concentration. Typically, a warning alarm will alert personnel that a threshold may be breached if action is not taken, allowing time to avoid an incident requiring a formal report. Critical alarms can alert operators that a threshold has been passed and automatically trigger the appropriate action, such as a systematic shutdown of the process.

Regulatory Agencies

Federal and national agencies maintain smaller state subsections of various programs and administrations. In most cases, the state level requires stricter compliance and lower limits. Solely state-controlled programs usually handle the air and water quality since any regulation violation results in the consequences for the immediate community.

Federal and National Agencies or Programs

Plant safety and environmental safety programs regulated on the national or federal level are monitored by three main agencies; the Environmental Protection Agency (EPA), the U.S. Department of Labor, the Food and Drug Administration, and the Department of Homeland Security. These three governing bodies have created numerous acts, committees, administrations and policies that protect the welfare of the employee, community, and environment.

The EPA at the federal level, provides acts, laws, and regulations, that help maintain and improve the air and water quality. The risk management program (RMP) is a mandatory program that *"require facilities that produce, handle, process, distribute, or store certain chemicals to develop a Risk Management Program"*. A Risk Management Plan (RMP) must be submitted to the EPA for approval. Overall, risk management is a large part of process control as control systems must adequately function and maintain compliance of an entire facility. Failure logic for instrumentation, redundant sensors, and critical alarms are essential in maintaining compliance, but a RMP is crucial for handling low-likelihood emergency situations.

The U.S. Department of Labor maintains the Occupational Safety and Health Administration (OSHA) which provides rules and regulations for employers and employees on safe workplace practices. Although individual states may maintain their own occupational health and safety plans, OSHA is the governing body and authority on those programs. Inspections are performed to ensure that all employees have a clean and safe working environment that is hazard-free and risk-mediated. OSHA also maintains the Process Safety Management Program (PSM) which regulates requirements for facilities that handle highly hazardous chemicals.

Process Safety Management (PSM)

Besides the catastrophic nature of events that can occur from neglecting Process Safety, large chemical facilities are granted a privilege and license to operate by the different federal regulatory agencies. If these regulatory agencies perform an audit on a specific facility, and find that their regulations are not being followed, then extremely large fines can be levied to the company, even to the extent of shutting the facility down permanently by removing that facility's privilege to operate.

In the case of PSM, it is OSHA who deals out these fines. For example, in 2009 OSHA attempted to levy a record *87 million* dollar fine to an integrated oil company, which has not been finalized in the legal system yet, but gives a good example of how important it is for companies operating in the U.S., if they want to continue to operate safely and economically, to follow all government regulations as closely as possible. Unexpected releases of toxic, reactive, or flammable liquids and gases in processes involving highly hazardous chemicals have been reported for many years in various industries that use chemicals with such properties. Regardless of the industry that uses these highly hazardous chemicals, there is a potential for an accidental release any time they are not properly controlled, creating the possibility of disaster.

As a result of catastrophic incidents in the past, and to help ensure safe and healthful workplaces, OSHA has issued the Process Safety Management of Highly Hazardous Chemicals standard (29 CFR 1910.119), which contains requirements for the management of hazards associated with processes using highly hazardous chemicals. OSHA's standard 29CFR 1910.119 emphasizes the management of hazards associated with highly hazardous chemicals and establishes a comprehensive management program that integrates technologies, procedures, and management practices.

An effective process safety management program requires a systematic approach to evaluating the whole process. Using this approach the process design, process technology, operational and maintenance activities and procedures, training programs, and other elements which impact the process are all considered in the evaluation. Process safety management is the proactive identification, evaluation and mitigation or prevention of chemical releases that could occur as a result of failures in process, procedures or equipment. OSHA prescribes essential tools to the success of process safety management including:

- Process Safety Information
- Process Hazard Analysis
- Operating Procedures and Practices
- Employee Training
- Pre-Startup Safety Review
- Mechanical Integrity
- Management of Change
- Incident Investigation
- Emergency Preparedness
- Compliance Audits

The thought is, with the simultaneous implementation of all of these things at a facility dealing with large amounts of highly hazardous chemicals, the risk of a catastrophic incident resulting from an unplanned release will be minimized. Following is a detailed discussion of each of these tools prescribed by OSHA.

PROCESS SAFETY INFORMATION(PSI)

Complete, accurate, and up-to-date written information concerning process chemicals, process technology, and process equipment is essential to an effective process safety management program. The compiled information will be a necessary resource to a variety of users including the team that will perform the process hazards analysis, those developing the training programs and operating procedures, contractors whose employees will be working with the process, those conducting the pre-startup safety reviews, local emergency preparedness planners, and insurance and enforcement officials. PSI includes, but is not limited to:

- Material and safety data sheets (MSDS)
- A block flow diagram showing the major process equipment and interconnecting process flow lines
- Process Flow Diagrams (PFDs)
- Piping and Instrument Diagrams (P&IDs)
- Process design information, including the codes and standards relied on to establish good engineering design

Process Hazards Analysis (PHA)

A process hazards analysis (PHA) is one of the most important elements of the process safety management program. A PHA is an organized and systematic effort to identify and analyze the significance of potential hazards associated with the processing and handling of highly hazardous chemicals. A PHA is directed toward analyzing potential causes and consequences of fires, explosions, releases of toxic or flammable chemicals, and major spills of hazardous chemicals. The PHA focuses on equipment, instrumentation, utilities, human actions, and external factors that might impact the process. These considerations assist in determining the hazards and potential failure points or failure modes in a process.

A team from each process unit in the facility will be tasked with conducting a PHA for their process unit at regularly scheduled intervals as defined by OSHA. One example is in an oil refinery, where a PHA has to be conducted and documented for each process unit every five calendar years. The competence of the team conducting the PHA is very important to its success. A PHA team can vary in size from two people to a number of people with varied operational and technical backgrounds.

The team leader needs to be fully knowledgeable in the proper implementation of the PHA methodology that is to be used and should be impartial in the evaluation. The other full or part time team members need to provide the team with expertise in areas such as process technology, process design, operating procedures and practices, alarms, emergency procedures, instrumentation, maintenance procedures, safety and health, and any other relevant subject as the need dictates. The ideal team will have an intimate knowledge of the standards, codes, specifications and regulations applicable to the process being studied.

There are various methodologies for conducting a PHA. Choosing which one is right for each individual facility will be influenced by many factors, including the amount of existing knowledge about the process.

Operating Procedures

Operating procedures provide specific instructions or details on what steps are to be taken or followed in carrying out the task at hand. The specific instructions should include the applicable safety precautions and appropriate information on safety implications. For example, the operating procedures addressing operating parameters will contain operating instructions about pressure limits, temperature ranges, flow rates, what to do when an upset condition occurs, what alarms and instruments are pertinent if an upset condition occurs, and other subjects. Another example of using operating instructions to properly implement operating procedures is in starting up or shutting down the process.

Operating procedures and instructions are important for training operating personnel. The operating procedures are often viewed as the standard operating practices (SOPs) for operations. Control room personnel and operating staff, in general, need to have a full understanding of operating procedures. In addition, operating procedures need to be changed when there is a change in the process. The consequences of operating procedure changes need to be fully evaluated and the information conveyed to the personnel.

For example, mechanical changes to the process made by the maintenance department (like changing a valve from steel to brass or other subtle changes) need to be evaluated to determine whether operating procedures and practices also need to be changed. All management of change actions must be coordinated and integrated with current operating procedures, and operating personnel must be alerted to the changes in procedures before the change is made. When the process is shut down to make a change, the operating procedures must be updated before re-starting the process.

Employee Training

All employees, including maintenance and contractor employees involved with highly hazardous chemicals, need to fully understand the safety and health hazards of the chemicals and processes they work with so they can protect themselves, their fellow employees, and the citizens of nearby communities.

Training conducted in compliance with the OSHA Hazard Communication standard will inform employees about the chemicals they work with and familiarize them with reading and understanding MSDSs. However, additional training in subjects such as operating procedures and safe work practices, emergency evacuation and response, safety procedures, routine and non-routine work authorization activities, and other areas pertinent to process safety and health need to be covered by the employer's training program.

In establishing their training programs, employers must clearly identify the employees to be trained, the subjects to be covered, and the goals and objectives they wish to achieve. The learning goals or objectives should be written in clear measurable terms before the training begins. These goals and objectives need to be tailored to each of the specific training modules or segments. Employers should describe the important actions and conditions under which the employee will demonstrate competence or knowledge as well as what are acceptable performance.

Careful consideration must be given to ensure that employees, including maintenance and contract employees, receive current and updated training. For example, if changes are made to a process, affected employees must be trained in the changes and understand the effects of the changes on their job tasks. Additionally, as already discussed, the evaluation of the employee's absorption of training will certainly determine the need for further training.

Pre-Startup Safety Review

For new processes, the employer will find a PHA helpful in improving the design and construction of the process from a reliability and quality point of view. The safe operation of the new process is enhanced by making use of the PHA recommendations before final installations are completed. P&IDs should be completed, the operating procedures put in place, and the operating staff trained to run the process, before startup. The initial startup procedures and normal operating procedures must be fully evaluated as part of the pre-startup review to ensure a safe transfer into the normal operating mode.

For existing processes that have been shut down for turnaround or modification, the employer must ensure that any changes other than "replacement in kind" made to the process during shutdown go through the management of change procedures. P&IDs will need to be updated, as necessary, as well as operating procedures and instructions. If the changes made to the process during shutdown are significant and affect the training program, then operating personnel as well as employees engaged in routine and non-routine work in the process area may need some refresher or additional training. Any incident investigation recommendations, compliance audits, or PHA recommendations need to be reviewed to see what affect they may have on the process before beginning the startup.

Mechanical Integrity

Employers must review their maintenance programs and schedules to see if there are areas where "breakdown" is used rather than the more preferable on-going mechanical integrity program. Equipment used to process, store, or handle highly hazardous chemicals has to be designed, constructed, installed, and maintained to minimize the risk of releases of such chemicals. This requires that a mechanical integrity program be in place to ensure the continued integrity of process equipment.

Elements of a mechanical integrity program include identifying and categorizing equipment and instrumentation, inspections and tests and their frequency; maintenance procedures; training of maintenance personnel; criteria for acceptable test results; documentation of test and inspection results; and documentation of manufacturer recommendations for equipment and instrumentation.

Management of Change

To properly manage changes to process chemicals, technology, equipment and facilities, one must define what is meant by change. In the process safety management standard, change includes all modifications to equipment, procedures, raw materials, and processing conditions other than "replacement in kind." These changes must be properly managed by identifying and reviewing them prior to implementing them.

For example, the operating procedures contain the operating parameters (pressure limits, temperature ranges, flow rates, *etc.* and the importance of operating within these limits. While the operator must have the flexibility to maintain safe operation within the established parameters, any operation outside of these parameters requires review and approval by a written management of change procedure. Management of change also covers changes in process technology and changes to equipment and instrumentation. Changes in process technology can result from changes in production rates, raw materials, experimentation, equipment unavailability, new equipment, new product development, change in catalysts, and changes in operating conditions to improve yield or quality.

Equipment changes can be in materials of construction, equipment specifications, piping pre-arrangements, experimental equipment, computer program revisions, and alarms and interlocks. Employers must establish means and methods to detect both technical and mechanical changes.

Temporary changes have caused a number of catastrophes over the years, and employers must establish ways to detect both temporary and permanent changes. It is important that a time limit for temporary changes be established and monitored since otherwise, without control, these changes may tend to become permanent. Temporary changes are subject to the management of change provisions.

In addition, the management of change procedures are used to ensure that the equipment and procedures are returned to their original or designed conditions at the end of the temporary change. Proper documentation and review of these changes are invaluable in ensuring that safety and health considerations are incorporated into operating procedures and processes. Employers may wish to develop a form or clearance sheet to facilitate the processing of changes through the management of change procedures.

A typical change form may include a description and the purpose of the change, the technical basis for the change, safety and health considerations, documentation of changes for the operating procedures, maintenance procedures, inspection and testing, P&IDs, electrical classification, training and communications, pre-startup inspection, duration (if a temporary change), approvals, and authorization. Where the impact of the change is minor and well understood, a check list reviewed by an authorized person, with proper communication to others who are affected, may suffice.

For a more complex or significant design change, however, a hazard evaluation procedure with approvals by operations, maintenance, and safety departments may be appropriate. Changes in documents such as P&IDs, raw materials, operating procedures, mechanical integrity programs, and electrical classifications should be noted so that these revisions can be made permanent when the drawings and procedure manuals are updated. Copies of process changes must be kept in an accessible location to ensure that design changes are available to operating personnel as well as to PHA team members when a PHA is being prepared or being updated.

Incident Investigation

Incident investigation is the process of identifying the underlying causes of incidents and implementing steps to prevent similar events from occurring. The intent of an incident investigation is for employers to learn from past experiences and thus avoid repeating past mistakes. The incidents OSHA expects employers to recognize and to investigate are the types of events that resulted in or could reasonably have resulted in a catastrophic release. These events are sometimes referred to as "near misses," meaning that a serious consequence did not occur, but could have.

Employers must develop in-house capability to investigate incidents that occur in their facilities. A team should be assembled by the employer and trained in the techniques of investigation including how to conduct interviews of witnesses, assemble needed documentation, and write reports. A multi-disciplinary team is better able to gather the facts of the event and to analyze them and develop plausible scenarios as to what happened, and why. Team members should be selected on the basis of their training, knowledge and ability to contribute to a team effort to fully investigate the incident.

Emergency Preparedness

Each employer must address what actions employees are to take when there is an unwanted release of highly hazardous chemicals. Emergency preparedness is the employer's third line of defense that will be relied on along with the second line of defense, which is to control the release of chemical. Control releases and emergency preparedness will take place when the first line of defense to operate and maintain the process and contain the chemicals fails to stop the release.

Employers will need to select how many different emergency preparedness or third lines of defense they plan to have, develop the necessary emergency plans and procedures, appropriately train employees in their emergency duties and responsibilities, and then implement these lines of defense. Employers, at a minimum, must have an emergency action plan that will facilitate the prompt evacuation of employees when there is an unwanted release of a highly hazardous chemical.

This means that the employer's plan will be activated by an alarm system to alert employees when to evacuate, and that employees who are physically impaired will have the necessary support and assistance to get them to a safe zone. The intent of these requirements is to alert and move employees quickly to a safe zone. Delaying alarms or confusing alarms are to be avoided. The use of process control centers or buildings as safe areas is discouraged. Recent catastrophes indicate that lives are lost in these structures because of their location and because they are not necessarily designed to withstand overpressures from shock waves resulting from explosions in the process area.

Preplanning for more serious releases is an important element in the employer's line of defense. When a serious release of a highly hazardous chemical occurs, the employer, through preplanning, will have determined in advance what actions employees are to take. The evacuation of the immediate release area and other areas, as necessary, would be accomplished under the emergency action plan. If the employer wishes to use plant personnel-such as a fire brigade, spill control team, a hazardous materials team-or employees to render aid to those in the immediate release area and to control or mitigate the incident, refer to OSHA's Hazardous Waste Operations and Emergency Response (HAZWOPER) standard.

If outside assistance is necessary, such as through mutual aid agreements between employers and local government emergency response organizations, these emergency responders are also covered by HAZWOPER. The safety and health protection required for emergency responders is the responsibility of their employers and of the on-scene incident commander.

Compliance Audits

An audit is a technique used to gather sufficient facts and information, including statistical information, to verify compliance with standards. Employers must select a trained individual or assemble a trained team to audit the process safety management system and program. A small process or plant may need only one knowledgeable person to conduct an audit. The audit includes an evaluation of the design and effectiveness of the process safety management system and a field inspection of the safety and health conditions and practices to verify that the employer's systems are effectively implemented.

The audit should be conducted or led by a person knowledgeable in audit techniques who is impartial towards the facility or area being audited. The essential elements of an audit program include planning, staffing, conducting the audit, evaluating hazards and deficiencies and taking corrective action, performing a follow-up review, and documenting actions taken.

Other Federal Entities

The Food and Drug Administration (FDA) is an agency of the United States Department of Health and Human Services. The FDA regulates food, drugs, cosmetics, biologics, medical devices, radiation-emitting devices and vetenary products manufactured in the United States. The main goal of the FDA is to maitain that the products they regulate are safe, effective, and secure. The FDA is also responsible that the products are accurately represented to the public. State and local governments also help regulate these products in cooperation with the FDA. The FDA does not regulate alcohol, illegal drugs, and meat and poultry.

The Department of Homeland Security has recently taken a role in regulating chemical plants because of 9/11. Chemical plants are seen by the government as targets for terrorists and security in and around the plant is a major concern. Although the laws are typically state run, the Department of Homeland Security has required mandatory national security standard to chemical plants throughout the nation. Although the mandates were fought by legislative for years, the Department of Homeland Security has influence in the security in chemical plants. The law requires the plant to prepare a vulnerability test and submit a site security plan. In order to validate these activities, audits and site visits are/will be performed by government officials.

Chapter 8

A SAW-Based Chemical Sensor for Detecting Sulfur-Containing Organophosphorus Compounds Using a Two-Step Self-Assembly and Molecular Imprinting Technology

Yong Pan[1,*], Liu Yang[1], Ning Mu[1], Shengyu Shao[1], Wen Wang[2], Xiao Xie[2] and Shitang He[2]

[1] State Key Laboratory of NBC Protection for Civilian, Research Institute of Chemical Defense, Yangfang, Changping District, Beijing 102205, China; E-Mails: Yangliujinjin@sina.com (L.Y.); Sdmuta@163.com (N.M.); Shaoshengyu@163.com (S.S.)

[2] Institute of Acoustic, Chinese Academy of Science, Zhongguancun Street, Haidian District, Beijing 100080, China; E-Mails: Wangwenwq@mail.ioa.ac.cn (W.W.); Xiexiao08@mails.ucas.ac.cn (X.X.); Heshitang@ mail.ioa.ac.cn (S.H.)

[*] Author to whom correspondence should be addressed; E-Mail: panyong71@sina.com.cn; Tel.: +86-136-6139-1971.

ABSTRACT

This paper presents a new effective approach for the sensitive film deposition of surface acoustic wave (SAW) chemical sensors for detecting organophosphorus compounds such as O-ethyl-S-2-diisopropylaminoethyl methylphosphonothiolate (VX) containing sulfur at extremely low concentrations. To improve the adsorptive efficiency, a two-step technology is proposed for the sensitive film preparation on the SAW delay line utilizing gold electrodes. First, mono[6-de-oxy-6-[(mercaptodecamethylene)thio]]-β-cyclodextrin is chosen as the sensitive material for VX detection, and a ~2 nm-thick monolayer is formed on the SAW delay line by the binding of Au-S. This material is then analyzed by atomic force microscopy (AFM). Second, the VX molecule is used as the template for molecular imprinting. The template is then removed by washing the delay line with ethanol and distilled water, thereby producing the sensitive and selective material for VX detection. The performance of the developed SAW sensor is evaluated, and results show high sensitivity, low detection limit, and good linearity within the VX concentration of 0.15–5.8 mg/m³. The possible interactions between the film and VX are further discussed.

Keywords

Surface acoustic wave (SAW); chemical sensor; self-assembled; molecularly imprinted (MIP); detection.

1. INTRODUCTION

Wohtjen [1] first reported the surface acoustic wave (SAW) method for gas sensing in 1979. Since then, SAW sensors have attracted great interest in the field of detection of poisonous and harmful gases at very low concentrations because of their small size, low cost, high sensitivity, and good reliability. The schematic and working principle of a SAW chemical sensor utilizing a dual-delay line oscillator structure is shown in Figure 1. SAW sensor responses arise from the mechanical interaction between the SAW and the sensitive film overlay. Thus, the adsorption efficiency of a sensitive film is a key factor affecting sensor performance, such as selectivity, sensitivity, stability, and response time. A SAW sensor responds to changes in mass on its surface with a shift in frequency and is thus most frequently used in gas-phase sensing applications. Accordingly, a chemically selective membrane should be used to collect and concentrate analyte molecules on the sensor surface by sorption. The sensitivity of a SAW sensor depends on the amount of vapor adsorbed and the inherent ability of the SAW transducer to respond to physical changes in the membrane caused by vapor adsorption. Therefore, we define sensitivity as the incremental change in signal occurring in response to an incremental change in analyte concentration, with detection limits depending on vapor sensitivity and on the noise of the sensor's signal.

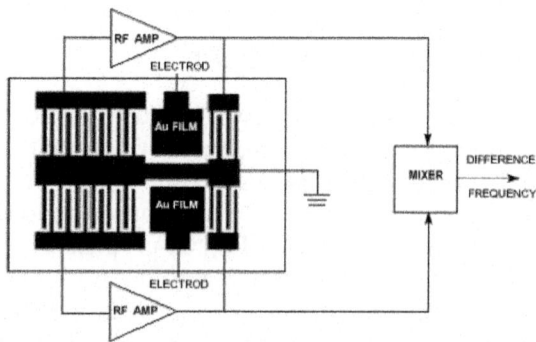

Figure 1. Schematic and principle of a SAW sensor.

Recognition layers are responsible for selective interactions with different types and forms of analytes. For instance, studies have shown polymers coated on SAW sensors are highly sensitive in qualitatively and quantitatively determining the composition of the gas [2], volatile organics [3], and chemical warfare agents [4-7]. For sensing purposes, SAW devices are coated with a molecular recognition or receptor layer, and adjoined to a frequency determining component. Depending upon the nature of a material, molecular recognition layers can be deposited onto the surface of SAW devices by spin coating, spray coating, sputtering, drop casting, electrospraying [8], or the Langmuir-Blodgett approach [9].

Self-assembled monolayer (SAM) technology provides an easy way of sensitive coating preparation with well-defined composition, structure, and thickness. Larry [10] reported, for the first time, a SAW chemical sensor for dimethyl methylphosphonate detection using SAW and self-assembled technology. Various substrates and functional groups have been investigated for SAM formation, including an alkanethiol monolayer on the Au surface. Although the growth mechanism of alkanethiol monolayers is unclear, alkanethiol SAMs are assumed to be formed as a result of chemical binding between the gold and sulfur atoms of thiol [11-13].

Molecular imprinting (MIP) technology has applications in syntheses, macromolecules, recognizability, and functional materials, among others [14]. The host-guest electronic, and steric complementarities, as well as host preorgnization, are three key elements in determining the stability of a complex in the gas phase [15]. By binding a monomer with a template molecule, a special three-dimensional structure could be formed after removing the template molecules in the sensitive coating. The target molecules could be successfully detected using this kind of selective and recognized coating [16,17]. Molecule host-guest systems, in which cyclodextrins (CDs) plays a very important role, are attracting increased interest. CDs are toroidal cyclic oligosaacharides with the secondary hydroxyls of glucose C-2 and C-3 on their more open face and the primary C-6 hydroxyl on the other face. The central cavities of CDs are hydrophobic or sterically restricted reaction fields; thus CDs are suitable host molecules. With a great variety of possible guests, this capacity has already been used in a number of applications [18,19]. β-CD is the most widely studied among the three CDs (α-, β-, γ-CD) because its cavity has the right size to bind a variety of aromatic and residues. In β-CD, the internal diameter of the cavity ranges from 6.0 to 6.5 Å, the average external diameter is about 15 Å, and the cavity depth is very close to 8 Å. As host-guest interactions can offer a novel approach to the functionalization of surface, thus, the chemisorbed layer of the host molecules offers a template upon which subsequent immobilization of guest molecules may occur. β-CD architecture, in which specific recognition reactions are used to build upon intricate structure in a controlled fashion, are possible by this method.

Although many technologies for sensitive film deposition have been successfully applied, most of them are inappropriate to prepare nanometer-thick films because of their poor film thickness, uniformity, stability, and reproducibility. Many advanced technologies such as LB require special equipment or are inadequate for certain polymers. In this paper, a simple way of preparing a self-assembled, molecularly imprinted coating for O-ethyl-S-2-diisopropylaminoethyl methylphosphonothiolate (VX) detection is proposed. First, mono[6-deoxy-6-[(mercaptodecamethylene)thio]]-β-CD is selected as the sensitive coating for VX detection and prepared on the gold surface of a SAW delay line by self-assembly. Second, VX molecule is used as a template for molecular imprinting. The developed SAW sensor using the new sensitive film deposition technique exhibits an excellent response to VX. The detection limit, selectivity, linearity and interference effect from the testing environment are experimentally evaluated.

2. EXPERIMENTAL SECTION

2.1. Reagents and Instruments

Mono[6-deoxy-6-[(mercaptodecamethylene)thio]]-β-CD was synthesized in our laboratory and analyzed by TOF-MS, FTIR, and NMR. O-Ethyl-S-2-diisopropylaminoethyl methylphosphonothiolate (Research Institute of Chemical Defence, Beijing, China). 1,10-decanedithiol (AR, TCI company, Tokyo, Japan), β-CD and mono(6-o-p-tolylsulfonyl)-β-CD (AR, Shanghai Chemical Reagent, Shanghai, China) and all other chemicals used were reagent grade.

As shown in Figure 1, dual SAW delay lines with separate frequencies were photolithographically fabricated on polished ST-quartz substrates. A 4 mm² Au film for sensing layer coating was then prepared between the interdigital transducers. Using the prepared SAW delay lines as the feedback element, two delay line oscillators were developed, and the sensitive film was deposited onto one device for gas sensing, with the naked device acting as the reference. The mixed differential oscillation frequency was used to characterize the target species, and recorded by a Model Proteck C3100 Frequency Counter (Proteck Company, Incheon, Korea). The vapor cell for detecting VX is made of aluminium and SAW dual lines are inserted in it, the cell is about 30 x 25 x 6 mm, there is an air hole on both sides of the cell so that VX could pass through by pumping, and the velocity of flow is 0.6 L/min.

2.2. Synthesis of Mono[6-deoxy-6-[(mercaptodecamethylene)thio]]-β-CD

The prepared β-CD was dried for 24 h at 100 °C in vacuum prior to utilization. Pyridine and N,N-dimethylformamide were distilled from CaH_2. Mono(6-o-p-tolylsulfonyl)-β-CD was prepared according to [20,21], and the crude product was recrystallized twice from acetone. Then, the product was reacted with 1,10-decanedithiol in an aqueous solution of Na_2CO_3 containing 20% ethanol at 50 °C to yield mono[6-deoxy-6-[(mercaptodecamethylene)thio]]-β-CD. The crude product was purified on a reversed-phase column. The synthesis route is depicted in Scheme 1.

Scheme 1. Synthesis of mono[6-deoxy-6-[(mercaptodecamethylene)thio]]-P-CD.

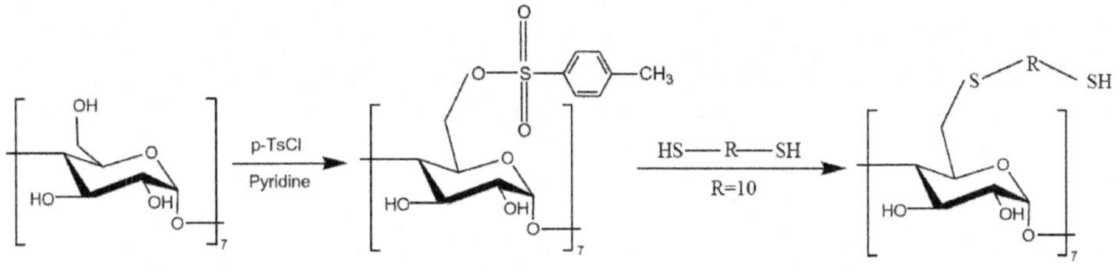

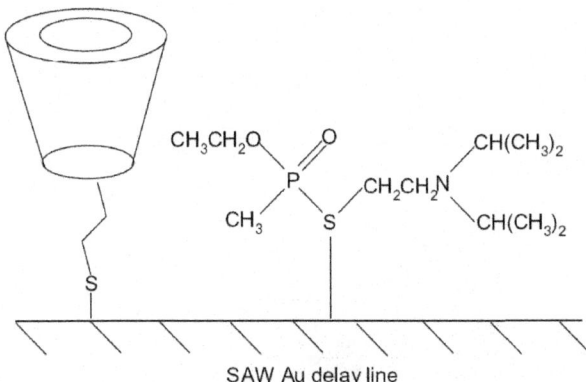

2.3. Preparation of a Self-Assembled, Molecularly Imprinted Film of SAW Sensor

Mono[6-deoxy-6-[(mercaptodecamethylene)thio]]-β-CD is a kind of alkanethiol, and VX is also a dialkyl sulfide. Both react very well on the Au surface to form Au-S bonds, so the Au surface is not entirely occupied by mono[6-deoxy-6-[(mercaptodecamethylene)thio]]-β-CD (Figure 2). Obviously, the self-assembled, molecularly imprinted film was not perfectly prepared.

SAW Au delay line

Figure 2. Competition between self-assembled film and molecular template on the surface of SAW Au delay line.

To solve this problem, we prepared a self-assembled, molecularly imprinted sensitive film by a two-step technology. The first step was the self-assembly of mono[6-deoxy-6-[(mercaptodecamethylene)thio]]-β-CD on the Au surface. The second step was molecular imprinting between the VX template and mono[6-deoxy-6-[(mercaptodecamethylene)thio]]-β-CD.

Prior to depositing the sensitive film, the Au surface of SAW delay lines were purged with a 3:1 (V/V), piranha (sulfuric acid, hydrogen peroxide) solution for cleaning before immersing into a 1×10^{-4} M mono[6-deoxy-6-[(mercaptodecamethylene)thio]]-β-CD ethanol solution for 24 h. Then, the SAW delay lines were washed with ethanol, and immersed into a 10 mM VX ethanol solution for another 24 h, VX may react with mono[6-deoxy-

6-[(mercaptodecamethylene)thio]]-β-CD in different binding ways, as shown in Figure 3a-c. After accomplishment of the binding between the template VX and mono[6-deoxy-6-[(mercaptodecamethylene)thio]]-β-CD, the SAW delay lines were washed with ethanol and distilled water to remove the VX templates. Consequently, the self-assembled, molecularly imprinted film was successfully formed on the Au surface of the SAW delay line.

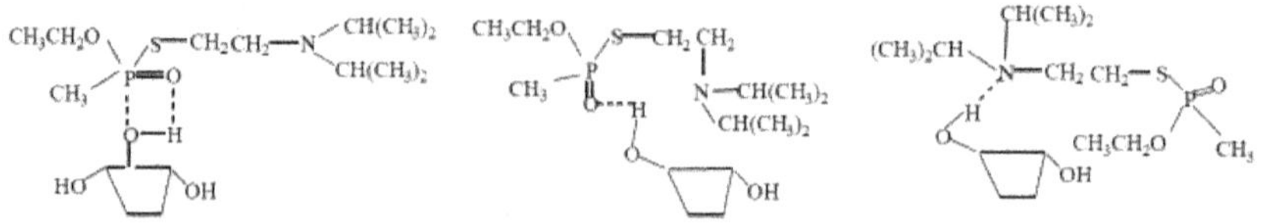

Figure 3. Reaction between self-assembled imprinted film and VX.

3. RESULTS AND DISCUSSION

3.1. Analysis by AFM and Calculations

AFM analysis was used to evaluate the covalent bonding and appearance of the self-assembled film surface, as shown in Figure 4. The RMS [Rq] of Au with no mono[6-deoxy-6-[(mercaptodecamethylene)thio]]-β-CD was only 2.294 nm, and this value increased to 12.014 nm when the Au delay line was coated. Obviously, mono[6-deoxy-6-[(mercaptodecamethylene)thio]]-β-CD was successfully formed on the Au surface of the SAW device.

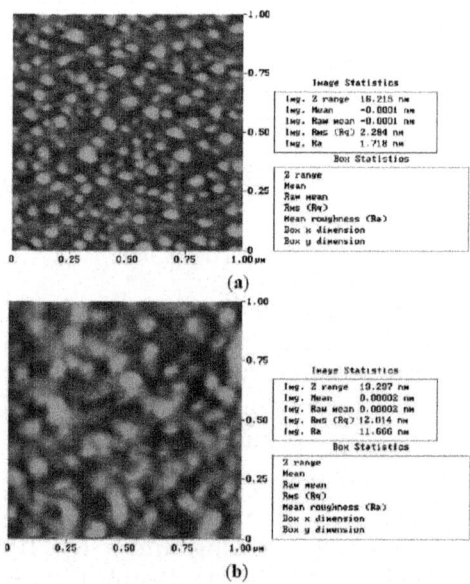

Figure 4. The 3D surface AFM photograph of SAW Au delay line. **(a)** Bare Au delay line, RMS [Rq] = 2.294 nm; **(b)** After self-assembled procedure, RMS [Rq] = 12.014 nm.

Given the high mass-sensitivity of the SAW sensor, the thickness of the self-assembled, molecularly imprinted film was estimated by measuring the frequency shift as follows [22,23]:

$$\Delta f = -1.26 \times 10^6 f_0^2 h \rho \qquad (1)$$

where Δf (Hz) is the frequency shift between coated and uncoated SAW delay lines. f_0 (MHz) is the operating frequency of the SAW sensor. h (cm) is the film thickness of the self-assembled, molecularly imprinted film, and ρ (g/cm^3) is the density of the film material.

The frequency of the SAW sensor was set to 300 MHz, the density of mono[6-deoxy-6-[(mercaptodecamethylene)thio]]-β-CD was 1.5 g/cm^3, and Δf was measured to be 10 kHz, so the thickness of SAM was estimated to be 2 nm on average.

3.2. Analysis of the MIP Effect

The MIP effect was investigated by comparing the response to VX of the MIP-coated sensor with that of the non-MIP-coated sensor, and the results are shown in Figure 5. When the developed non-MIP coated SAW sensor was exposed to 1.5 mg/m^3 VX, a very weak frequency response of only 2 kHz was observed because of the lack of particular 3D space structures. After MIP, as the molecular imprinting approached the covalent process, the covalent imprinting mainly depended on an easily cleavable arrangement between the template and the monomeric compound, which induced cavity formation. Given the imprinting effect, when VX was detected using the MIP-coated sensor, about 7 kHz frequency shift was detected. Moreover, a larger frequency response was obtained over the non-MIP-coated SAW sensor, hence, hence, the MIP effect was experimentally confirmed.

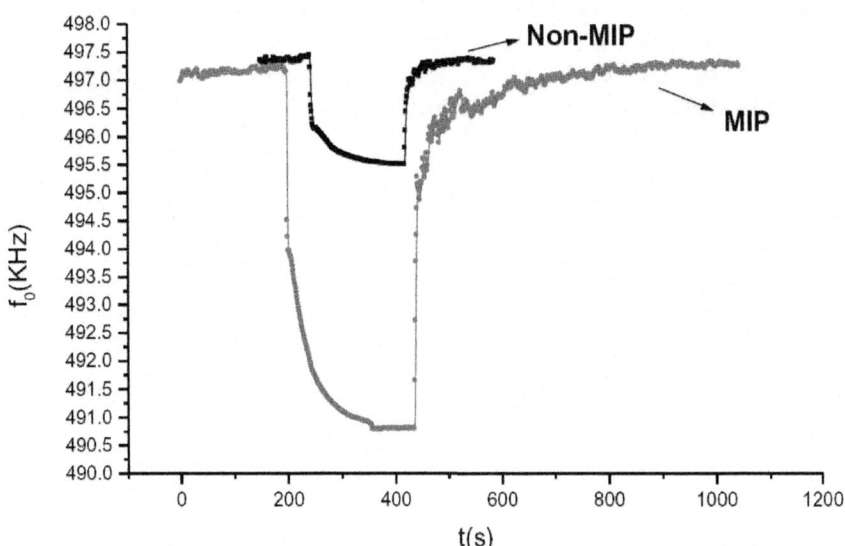

Figure 5. Confirmation of MIP effect (28 °C , RH = 70%).

3.3. Sensor Response to VX Detection

Table 1 shows that VX was detected under different test conditions. With decreased VX concentration, the sensor frequency decreased and a longer response time was observed. At high concentrations, much more VX molecules in the unit gas phase were adsorbed by the mono[6-deoxy-6-[(mercaptodecamethylene)thio]]-β-CD coating, so the time to reach equilibrium was shortened. On the contrary, at low concentrations, lower gas molecule adsorption led to a relatively low sensor response. Establishment of the response equilibration also took a long time, which resulted in a long response time. According to IUPAC methods, the detection limit toward VX was evaluated as 0.15 mg/m^3 (S/N > 3), and the linearity response to VX ranged within 0.15–5.8 mg/m^3.

Table 1. Response of different concentrations to VX (20 ° C , RH = 50%).

Concentration (mg/m^3)	Frequency Shifts (Hz)	Response Time (min)	Recovery Time (min)
5.80	2850	1.7	4.5
4.30	2460	1.9	4.3
2.67	1715	2.3	4.1
1.60	1507	2.4	3.5
0.85	1120	2.8	2.9
0.55	860	5.6	2.3
0.15	437	8.5	1.1

To study the stability of this sensor system, the MIP sensor was kept at room temperature, the typical response curve for four successive VX exposures is shown in Figure 6. And VX was detected with it under the same condition for many months (Figure 7). In the first 6 months, the detection signal for VX decreased by about 3.3%, after 18 months, the decrease reached about 4.4% and then abated, so VX could still be detected with the MIP sensor.

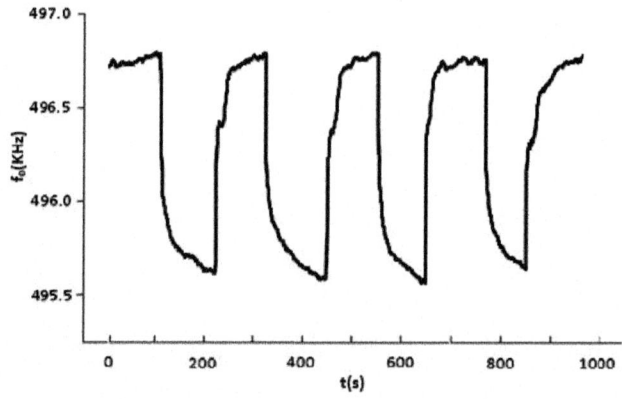

Figure 6. Response curve for four successive exposures of the sensor to VX at 1.25 mg/m^3.

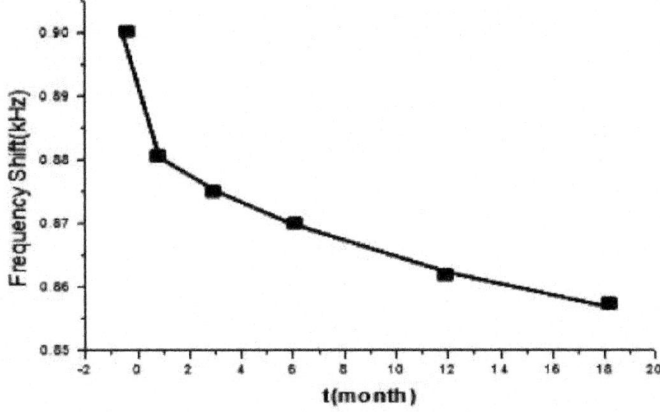

Figure 7. Stability study of MIP sensor.

3.4. Anti-Interference Experiment of the SAW-MIP Sensor

To further confirm the MIP effect, many other gases with larger concentrations (100–1000 times that of VX) were used to perform an anti-interference experiment. The results are listed in Table 2. Almost no influence on the sensor response was observed from most common organic solvents and gases. However, organic amines and organic acids at high concentrations obviously affected the sensor because they exhibited extend binding or adsorptivity on the surface of sensitive film. Nevertheless, the resulting frequency shifts were much lower than that of VX.

Table 2. Response of the SAW-MIP sensor to interferences.

Interference Gas	Concentration (mg/m³)	Frequency Shifts (Hz)
Omethoate	1,000	1,123
CH_3OH	10,000	168
CH_3CH_2OH	10,000	123
HCOOH	1,000	420
CH_3COOH	1,000	360
$CH_3(CH_2)_4COOH$	1,000	1,360
NH_3	2,000	403
$C_6H_5NH_2$	2,000	516
O-Anisidine	1,000	212
$C_2H_5OC_2H_5$	10,000	103
Petrdeumether	10,000	169
THF	10,000	230
n-C_6H_{14}	10,000	197

n-C_8H_{18}	1,000	341
CCl_4	10,000	214
HCHO	1,000	125
CH_3COCH_3	10,000	118
$CH_3COOC_2H_5$	10,000	103
C_6H_6	10,000	189
$C_6H_5CH_3$	1,000	226
C_6H_5Cl	1,000	103
H_2O	10,000	2,213
CH_3CN	1,000	205
Smog	high	-

4. CONCLUSIONS

The physicochemical properties of a chemoselective material are very critical to performance improvement in chemical sensing applications. In particular, the chemical preparation technique for sensitive films affects the coating uniformity, adhesion, and quality of the sensor. So many chemoselective coatings for functionalized self-assembled monolayer structures have been utilized to develop SAW chemical sensors that can exhibit a very rapid response at extremely low concentrations.

In this work, a novel two-step self-assembly and molecular imprinting technology for preparing sensitive films used to detect warfare agents VX was developed. The technology produced a film containing covalently immobilized onto host molecules (mono[6-deoxy-6-[(mercaptodecamethylene)thio]]-β-CD). The SAW sensor coated with the self-assembled, molecularly imprinted film was very sensitive to VX, and a significant frequency response was observed. High sensitivity and excellent selectivity were obtained because of the host-guest interaction between mono[6-deoxy-6-[(mercaptodecamethylene)thio]]-β-CD cavity and VX.

Overall, our results indicated that mono[6-deoxy-6-[(mercaptodecamethylene)thio]]-β-CD with good chemical selectivity can be successfully designed using appropriate molecules for the formation of self-assembled, molecularly imprinted films on SAW gold delay line. Related works are currently in progress in our laboratory.

Acknowledgments

This work was partly supported by National Natural Science Foundation of China (No.11074288).

Author Contributions

All authors participated in the work presented here. Yong Pan defined the research topic, he wrote the paper with Wen Wang. Liu Yang and Shengyu Shao carried out most of the experiments. Ning Mu and Xiao Xie contributed to the experimental measurements. Shitang He reviewed and edited the manuscript. All authors read and approved the manuscript.

Conflicts of Interest

The authors hereby declare no conflict of interest.

REFERENCES

1. Wohltjen, H.; Dessy, R. Surface acoustic wave probe for chemical analysis. *Anal. Chem.* **1979,** *51,* 1458–1475.

2. Al-Mashat, L.; Wlodarski, W.; Kaner, R.B.; Kalantar-zadeh, K. Polypyrrole nanofiler surface acousticv wave gas sensors. *Sens. Actuators B Chem.* **2008,** *134,* 826–831.

3. Park, Y.; Dong, K.-Y.; Lee, J.; Choi, J. Development of an ozone gas sensor using single-wassed carbon nanotubes. *Sens. Actuators B Chem.* **2009,** *140,* 407–411.

4. Tasaltin, C.; Gurol, I.; Harbeck, M.; Musluoglu, E. Synthesis and DMMP sensing properties of fluoroalkyloxy and fluoroaryloxy substuted phthalocyanines in acoustic sensors. *Sens. Actuators B Chem.* **2010,** *150,* 781–787.

5. Joo, B.-S.; Huh, J.; Lee, D. Fabrication of polymer SAW sensor array to calssfy chemical warfare agents. *Sens. Actuators B Chem.* **2007,** *121,* 47–53.

6. Matatagui, D.; Marti, J.; Fernandez, M.J. Chemical warfare agents simulants detection with an optimized SAW sensor array. *Sens. Actuators B Chem.* **2011,** *154,* 199–205.

7. Matatagui, D.; Fernandez, M.J.; Forntecha, J. Love-wave sensor array to detect, discriminate and classify chemical warfare agent simulants. *Sens. Actuators B Chem.* **2012,** *175,* 173–178.

8. Ganan-Calvo, A.M.; Davila, J.; Barrero, A. Current and droplet size in the electrospraying of liquids-Scaling laws. *J. Aerosol Sci.* **1997,** *28,* 249–275.

9. Petty, M.C. Possible applications for Langmuir Blodgett films. *Thin Solid Films* **1992,** *210–211,* 417–426.

10. Kepley, L.J.; Crook, R.M. Selective surface acoustic wave-based organophonate chemical sensor employing a self-assembled composite Monolayer: A new paradigm for sensor design. *Anal. Chem.* **1992,** *64,* 3191–3193.

11. Kim, J.; Park, S. Analysis of Im-SH self-assembled monolayer formation and its interaction with Fe^{2+} and Zn^{2+} using quartz chemical analyzer. *Sens. Actuators B Chem.* **2001,** *76,* 74–79.

12. Ulman, A. Formation and structure of self-assembled monolayerls. *Chem. Rev.* **1996,** *96,* 1533–1554.

13. Hhomas, H.; Hubert, K.; Otto, S.W. A simple strategy for preparation of sensor arrays: Molecularly structured monolayers as recognition elements. *Chem. Commun.* **2003,** 432–433.

14. Dickert, F.L.; Haunschild, A.; Kuschow, V.; Reif, M.; Stathopulos, H. Mass-sensitive detection of solvent vapors-mechastic studies on host-guest sensor principles by FT-IR spectroscopy and BET adsorption anslysis. *Anal. Chem.* **1996,** *68,* 1058–1061.

15. Dickert, F.L.; Bäumler, U.P.A.; Stathopulos, H. Mass-sensitive solvent vapor detection with calyx[4]-resorcinarenes: Turning sensitive and predicting sensor effects. *Anal. Chem.* **1997,** *69,* 1000–1005.

16. Yang, X. Chemical Microsensors for Detection of Explosives and Chemical Warfare Agents. U.S. Patent 6,316,268, 28 May 1998.

17. Elena, G.; Norbert, M.; Karsten, H. Analyte templating: Enhancing the enantioselectivity of chiral selectors upon incorporation into organic polymer environments. *Anal. Chem.* **2005,** *77,* 5009-5018.

18. Yang, X.; Swanson, B. Molecular host siloxane thin films for surface acoustic wave chemical sensors. *Sens. Actuators B Chem.* **1997**, *45*, 79-84.

19. Wang, C.H.; Xu, X.L. Electrochemical investigation of parathion impringred sensor and its application. *Chin. J. Appl. Chem.* **2006**, *23*, 404-408.

20. Keiko, T.; Kenjiro, H. Monotosylated a- and p-cyclodextrins prepared in an alkaline aqueous solution. *Tetrahedron Lett.* **1984**, *25*, 3331-3334.

21. Akihiko, U.; Ronald, B. Selective sulfonation of a secondary hydroxyl group of (3-cyclodextrin. *Tetrahedron Lett.* **1982**, *23*, 3451-3454.

22. Franz, L.; Peter, L. Molecular imprinting in chemical sensing detection of aromatic and halogenated hydrocarbons as well as polar solvent vapors. *Fresenius J. Anal. Chem.* **1998**, *360*, 759-762.

23. Cao, B.Q.; Huang, Q.B. Study on the surface acoustic wave sensor with self-assembled imprinted film of calixarene derivatives to detect organophosphorus compounds. *Am. J. Anal. Chem.* **2012**, *3*, 664-668.

INDEX